NEOFINALISM

Cary Wolfe, Series Editor

(continued on page 301)

NEOFINALISM

RAYMOND RUYER

Translated by Alyosha Edlebi

Introduction by Mark B. N. Hansen

posthumanities 36

UNIVERSITY OF MINNESOTA PRESS
Minneapolis · London

Originally published in French in 1952 as *Néo-finalisme*; copyright Presses Universitaires de France, second edition 2012.

Published by the University of Minnesota Press
111 Third Avenue South, Suite 290
Minneapolis, MN 55401-2520
http://www.upress.umn.edu

Library of Congress Cataloging-in-Publication Data

Ruyer, Raymond, 1902–1987.
Neofinalism / Raymond Ruyer ; translated by Alyosha Edlebi ; introduction by Mark B. N. Hansen.
Includes bibliographical references and index.
ISBN 978-0-8166-9204-0 (hc)
ISBN 978-0-8166-9205-7 (pb)
1. Teleology. 2. Philosophy, Modern. I. Title.
BD542.R8713 2016
124—dc23 2015019116

Printed in the United States of America on acid-free paper

21 20 19 18 17 16 10 9 8 7 6 5 4 3 2 1

Contents

Introduction

Form and Phenomenon in Raymond Ruyer's Philosophy

MARK B. N. HANSEN

> Philosophy has a duty to avoid snobbery, but also not to fall prey to timidity.
>
> **RAYMOND RUYER, "L'esprit philosophique"**

Raymond Ruyer (1902–87) was born in Plainfaing in the Department of Vosges in the Lorraine region of northeastern France. A precocious student who at nineteen received a first on his college entrance exam, Ruyer pursued a course of study in philosophy at the prestigious École Normale Supérieure and aggregated in 1924. While teaching at the lycée of Saint-Brieuc, Ruyer published his two theses, *Esquisse d'une philosophie de la structure* (Outline of a philosophy of structure) and *L'Humanité de l'avenir d'après Cournot* (Humanity of the future according to Cournot). In 1934, he returned to the Vosges region, taking up a position at the University of Nancy, where he was subsequently appointed *maître de conférences* (1939) and professor (following the war). From 1940 to 1945, Ruyer was a prisoner of war in a camp for French soldiers in Edelbach, Austria, where he participated in a vibrant intellectual culture with a number of scholars, including the biologist Étienne Wolff and the geologist François Ellenberger, both of whom would go on to prominence. It was during this time of internment that Ruyer wrote what would become the first exposé of his mature system, *Éléments de psycho-biologie* (Elements of psychobiology; published in 1946). After the war, Ruyer would go on to a solid university career as professor of philosophy at the University of Nancy. Author of twenty-two books and more than one hundred articles, Ruyer fully embraced his penchant for philosophy of the *esprit métaphysique,* by which he meant philosophy in its "proper form," philosophy that "is interested" not only in "everything [*au tout*]" but "in Totality [*au Tout*]."[1]

The publication of the English translation of Ruyer's masterpiece,

Neo-Finalisme (1952), will make a crucial contribution to recent activity aimed at redirecting attention to this unduly neglected twentieth-century philosopher. The critical neglect of Ruyer's thought is by no means exclusive to English-language circles, whose exposure has until now been limited to a few articles published in the interdisciplinary philosophy journal *Diogenes*.[2] Indeed, as Fabrice Colonna observes in his editorial presentation for a special issue of the French journal *Les Études philosophiques*, devoted to Ruyer, "the fact is that Ruyer has been purely and simply effaced from the field of theoretical references."[3] This fact is all the more unfortunate given the manifold resonances that link Ruyer's philosophical corpus to key concerns of contemporary philosophy and cultural theory in both English- and French-language circles, including such hot-button issues as the mind–body problem or the presuppositions behind biological morphogenesis.

More than any other factor, it is Ruyer's attitude toward science that may ultimately account for whatever impact his work will have on contemporary intellectual debates. Like William James and Alfred North Whitehead, both of whom are currently undergoing revitalizations of their own, and like his compatriot Gilbert Simondon, whose work has for some time now garnered critical attention and respect, Ruyer is a philosopher who not only deeply respects the work of scientists but believes at heart that his own work bears a crucial responsibility to science. The ultimate aim of Ruyer's philosophy is precisely to provide the metaphysical basis on which the empirical findings of scientific research can be made to cohere.

THE "DUALITY" OF MIND AND APPEARANCE

This attitude is clearly manifest almost from the very outset of Ruyer's career. One could argue that Ruyer's break with his own initial devotion to structuralism—a devotion that finds quintessential expression in his primary thesis *(Esquisse d'une philosophie de la structure)*—represents the single most consequential development of his entire philosophical career. For it is this break, and the differentiation of functioning *(fonctionnement)* from structure, that informs Ruyer's later theorization of a "transspatial" domain of form that lies beneath and in-forms what happens in space and time. As we shall see, this key insight stems from

Ruyer's philosophical working-through of scientific research, in this case, in the burgeoning field of embryology: not only does Ruyer aim to explain how the embryo holds the potential for the future development of an organism but he seeks to develop the process of embryogenesis *into a philosophical model for the operation of form as such*.

The first step in—and in a certain sense, the theoretical cornerstone of—Ruyer's philosophical transfiguration of science is his meditation on consciousness in his 1937 book *La conscience et le corps* (Consciousness and the body). In this work, his first post-thesis book-length publication, Ruyer argues (somewhat scandalously to me upon my initial reading of it some fifteen years ago) that consciousness is an absolute form of being and, as a consequence, that the body is epiphenomenal. Yet far from seeking to displace the body, Ruyer's aim here is to overcome the dualism between consciousness and matter that, in his opinion (and on this point, he is certainly not alone), has plagued the history of modern philosophy from Descartes onward. Ruyer's point, it is important to emphasize, is a methodological one: the consciousness–matter divide raises what, for him, is simply a badly posed question. That is why, in the place of the mind–brain dualism central to the mind–body problematic that has informed Western philosophy from Descartes to contemporary analytic philosophy, Ruyer introduces a different kind of dualism—a "duality" between mind and appearance.

Informing this duality is Ruyer's conviction that consciousness, to be what it is, must possess itself absolutely. Consciousness must, that is, enjoy a sensory reality prior to and independent of any perceptual reality it may give rise to or that may be given to it. In this sense, Ruyer's philosophy inaugurates a line of exploration that diverges from the dominant strains of philosophy, and especially of phenomenology, in vogue at the time of his apprenticeship and in the early phases of his academic career. Where phenomenology generically takes intentionality, the relation of consciousness to an object or the "aboutness" of consciousness, as a primitive, Ruyer's philosophy of consciousness insists on absolute sensation as its foundation. Consciousness does not *have* a visual (or phenomenal) field as its intentional object. It *is* this field itself.

Ruyer's position entails the valuation of unity over against distance: consciousness is above all absolute self-possession or unity, and the absolute sensation composing its being is simply *without distance*. Indeed,

Ruyer's own departure from phenomenology is concretely marked by his criticism of perception for its tendency to distract us from, and indeed actively to make us forget, this determining condition of consciousness. Because we characteristically see the world as an image standing in front of us, and because we see parts of our body as elements in such an image, we naturally tend to think of our visual field as being something out in front of us, at a distance from our sensation. This habitual state of affairs easily leads to confusion when we mistake what is in reality a derivative of absolute sensation, the product of motor activity rooted in that sensation, for the source of sensation itself.

In opposing this illusion of perception, Ruyer recurs to the position of Bishop Berkeley, whose 1709 *Essay towards a New Theory of Vision* established that distance is not given by vision and that vision is primordially without any dimension of depth. What Ruyer adds to Berkeley's position is consideration of a richer body of scientific research, including work on retinal disparity, eye movement, and object recognition, as well as a cleaner separation of philosophical from empirical perspectives. From Ruyer's perspective, the error that is practically unavoidable in normal life—an error owing to the purely contingent fact that we literally see parts of ourselves in the visual field—stems from the confusion between, indeed, the conflation of, the mechanics of vision and the mode of being of sensation. Though science can tell us much about the former, it is only philosophy, indeed, philosophy of the metaphysical variety, meditating on the empirical findings of science, that can properly address absolute sensation in its primordial being.

Ruyer's conception of the absolute self-possession of consciousness and his critique of the illusion of perception provide a quintessential example of his philosophical method, illustrating precisely how it philosophically transfigures the findings of science in the service of metaphysics. Yet Ruyer's methodological transfiguration isn't limited to what it tells us about the philosophical significance of the findings of science, and Ruyer cannot in any way be pigeon-holed as a philosopher of science. Rather, Ruyer's aim—and the motivation for his recourse to science in the first place—is to develop a philosophy capable of capitalizing on scientific research to advance the project of philosophy, understood as a separate domain of inquiry autonomous from and more general than science. On this score, the achievement of Ruyer's

La conscience et le corps is less its philosophical evaluation of research in science than the model of reality it builds from it, which posits absolute consciousnesses as primary reality and views bodies and objects, as well as other consciousnesses, as appearances of and to such reality.

By defining other minds in this way, as the appearance of other absolute realities to the absolute reality that is my mind, Ruyer both displaces mind–brain dualism in favor of a duality between mind and appearance and relativizes dualism itself to the point of transforming it into what Fabrice Colonna has called "spiritual monism."[4] For Ruyer, there is "only one single reality," that of mind or spirit *(esprit)*, which presents "two aspects": "true being" and "a phenomenon."[5] As against phenomenology, where reality and phenomenon address one and the same object (albeit in vastly different ways), and where the phenomenon, specifically, is the appearance (technically, the "adumbration") of the object, for Ruyer, reality and phenomenon diverge fundamentally. Far from being the appearance of an underlying reality, phenomenon on Ruyer's account designates the mode in which all temporally and spatially extended reality is experienced. As such, the phenomenon does not and cannot share an object in common with absolute sensation but is made of a wholly separate, though derivative, domain of existence. If absolute consciousness is simply absolute sensation, sensation without any phenomenal dimension whatsoever, the phenomenon, despite finding its source in absolute sensation, is an achievement in its own right, even—or indeed, especially—in the form of the appearance of other absolute consciousnesses.

It is precisely this disjunction between the reality of sensation and the phenomenal domain of appearance that makes Ruyer's philosophy of such great relevance for contemporary debates concerning neuroscience and consciousness. For with his "spiritual monism," Ruyer needn't choose between mind and matter, and with his relativizing functionalization of the consciousness–appearance duality, he can avoid placing mind and brain on the same flat continuum (a continuum that eventuates in eliminative materialism). Indeed, by integrating both absolute sensation and phenomenal appearance into a single, coherent, functionally differentiated account, Ruyer is able to do what phenomenological approaches and analytic philosophy have failed to do, namely, to give *both* parties their due. Indeed, Ruyer's understanding of the coupling of absolute consciousness and phenomenal appearance addresses precisely what goes missing in most other accounts: the fact that brain activity is the basis for mental phenomena, even as the latter, as distinct activity in their own right, are not reducible to the former. In the words of Fabrice Colonna, Ruyer's spiritual monism "is the only philosophy that makes it possible to account at one and the same time for the perspective [*témoignage*] of consciousness and for the findings [*données*] of neurology."[6] In this sense, it anticipates—and perhaps already advances beyond—recent work in neuropsychology (Mark Solms) and neuroplasticity (Catherine Malabou).

TRUE FORM AND SECONDARY OBJECTS

Ruyer erects his mature philosophy, developed in *Éléments de psychobiologie* (1946) and in his masterwork, *Neo-Finalisme,* on the foundation provided by this functional differentiation between absolute sensation and phenomenal appearance. *Éléments* marks Ruyer's definitive break with his earlier embrace of structure, which he now characterizes as nothing more than a "symptom."[7] Once again, Ruyer's method is to work through scientific research, this time in the field of morphogenesis, to flesh out a key philosophical argument concerning form that will be central for the rest of his career. Taking on one of the heroic forefathers of contemporary complexity theory, D'Arcy Wentworth Thompson, and developing an argument that would apply with equal force to the work of Stuart Kauffman or Brian Goodwin, Ruyer contests the viability of theories of emergence that claim to explain complex forms, for example, organs, as products of the complexification of simple dynamic actions. According to Ruyer, such theories of emergence can only explain the behavior of what he calls "crowds," inanimate objects that are entirely without consistency. For Ruyer, that is, emergent properties are crowd phenomena that, like the appearance of other minds in the account of *La conscience et le corps,* are derivative from and have no power whatsoever to explicate the absolute unities composing them. From Ruyer's perspective, in short, theories of emergence cannot account for what is important, namely, the existence of absolute unities. Moreover, because they are unable to differentiate crowds from unities, such theoretical understandings, like perceptual consciousness in its everyday functioning, can only remain ignorant of their own incapacity.

On the basis of his critique of emergence in *Éléments,* Ruyer develops a distinction between "primary being" and "secondary object" that sheds light on the specificity of form in his conceptualization of it. Once again, Ruyer's development marks an unequivocal turn away from the concept of structure: for, because both primary being and secondary object have a structure that "is the result of a dynamism," what differentiates them must lie elsewhere, namely, in the "nature" of this dynamism. In cases where the forces that constitute a dynamic system are "simply added together" or "composed according to simple laws of vectorial calculus," it is a secondary object that is at issue. Only when such forces constitute a system that is "coordinated, individualized, and subsistent through time" can we properly speak of the operation of a form.[8]

Ruyer specifically distinguishes his concept of form, or "form in itself," from the concept central to Gestalt psychology: form in itself, or "true form," implies, "in contrast to the form-*gestalt,* not only instantaneous unity but unity that is dominant through the time of successive states."[9] This self-producing and self-sustaining aspect is precisely what defines form in itself: "a form," explains Ruyer, "exists in itself, as such. . . . Being in itself, a form is thus its own subject."[10] It is "dynamically individualized and self-subsisting."[11] Rejecting the philosophical, and specifically Sartrean, distinction between the *in itself* and the *for itself,* Ruyer contends that all existence is necessarily both in and for itself. Put more contentiously, this is to say that "only primary beings"—or true forms—"exist" and, by implication, that secondary phenomena and objects "only represent the set of equilibrated interactions of primary beings."[12]

Ruyer's conceptualization of "form in itself" marks an important development in relation to his earlier account of absolute consciousness for two reasons. First, with its specification of how dynamic structure must operate in the case of primary being, that is, as temporally subsistent, form necessarily broaches the crucial notion of potential. A true form can never be static, can never be a pure actuality, precisely because of its operation to shape the development of structure through time. Form, or the "formal theme" of an organism, is *potential* "insofar as it commands not just an instantaneous structure but the coordinated succession of structures that appear to us in time."[13] The development of human anatomy, for example, of which the adult state is only a "cross section" or "cut [*coupe*]," occurs in relation to a potential that is never fully actualized by any particular phase. Ruyer grasps the metaphysical significance of potentiality when he specifies that its operation *does not occur in physical space-time.* Potential, he contends, is "outside" of the space-time of the physicist, and indeed, it is the very passage of potential into space-time that organizes structure.

If this thematization of the crucial operation of potential anticipates Ruyer's central concept of the transspatial and transtemporal domain, a concept that lies at the heart of *Neofinalism,* it also serves to express, in yet another configuration, Ruyer's distinctive metaphysical embrace of science. For in claiming that potential lies outside physical space-time, Ruyer makes common cause—as he himself explicitly notes—with quantum physics. A form or formal theme can be likened to a physical atom in that the "quantified character of action" in it, like that in the atom, "is the indication that something outside of space-time organizes what appears in space-time as an indecomposable whole [*ensemble*]. The quantified time of action cannot be time as the pure dimension of macroscopic physics."[14] Here we find evidence of Ruyer's extensive indebtedness to quantum physics—for him, quantum nonlocality furnishes a key example of the transspatial and transtemporal character of absolute forms—but also, and of equal importance, we come upon his generalization of absolute sensation beyond the domain of living beings.

With this crucial development, we encounter the second key contribution of Ruyer's concept of form: the domain of true form, or form in itself, extends across the entire expanse of the material universe. Far from being restricted to beings of a certain complexity, there are true forms at every scale of physical reality. What this means is that Ruyer, like Whitehead before him, extends subjectivity to entities of all scales. In his own explanation of this development, Fabrice Colonna invokes the key concept of the "immanent" or "absolute survey" *(survol absolu)* that, though introduced in *Éléments,* will be given its full due only in the amazing account Ruyer gives in chapter 9 of *Neofinalism.* According to Colonna's explanation, what microphysics introduces—or more precisely, what Ruyer's metaphysical transfiguration of microphysics exposes—is the capacity of matter to survey itself: "with the advent of [quantum] microphysics, the challenge to matter's traditional properties has made the notion of absolute domain applicable to matter itself, on account of the character of delocalized unity through immanent survey that one finds equally in the conscious brain and in the microphysical entity."[15]

This transfiguration of microphysics notwithstanding, it remains the case that Ruyer focuses his major philosophical investment on developing a philosophy of life capable of escaping the impasses of both vitalism and mechanistic materialism. Not only does Ruyer choose to emphasize the biological domain of embryology rather than the physical domain of quantum phenomena but the entire trajectory leading to his final conceptualization of the "absolute survey" proceeds in relation to the operation of consciousness. Whether he develops it in relation to what he calls "psychism" *(Éléments)* or in relation to consciousness proper *(Neofinalism)*, Ruyer introduces a distinction between primary form and secondary form that is absolutely definitive for his philosophical approach to life. In contrast to the overwhelming tendency of Western philosophical texts to privilege the higher-order operations of human consciousness, Ruyer insists that primary psychism or consciousness is fundamental, with secondary psychism or consciousness being either a subordinate development from it or, in the most radical formulation, an appearance of this sole primary form.

To understand the significance of Ruyer's philosophy, and of his articulation of a new holist and (in a sense to be specified) "theological" form of finalism, it is absolutely imperative that we grasp the fundamental—and fundamentally counterintuitive—gesture introduced by his privileging of primary consciousness. Independently of the question of their respective complexity, primary consciousness can be distinguished from secondary consciousness on account of its self-relatedness. As a basal self-relation—or self-enjoyment—upon which everything else is built, primary consciousness furnishes the basis *for all forms of consciousness,* primary and secondary alike. Colonna captures this foundational dimension perfectly when he observes that "all consciousness is first of all centered on itself, and only sees itself."[16]

This foundational dimension has important consequences for how we understand the relationship of primary and secondary consciousness. Far from differing primarily or solely on account of their respective degrees of complexity and their objects, these two forms of consciousness designate what are, ultimately, entirely distinct operations. Whereas secondary consciousness centers on how the world is represented by and for primary consciousness, primary consciousness itself is concerned exclusively with consciousness's capacity directly to live itself—with consciousness *as absolute survey.*

In his own effort to clarify this absolutely crucial operational distinction, Ruyer differentiates consciousness as absolute form or being from consciousness as "knowledge-correspondence." This distinction, which locates Ruyer's thinking at a right angle in relation to neuroscience, speaks directly to his philosophical affinities with and differences from phenomenology: "consciousness is not 'knowledge of' . . . , it is reality, it is 'being.' Consciousness becomes knowledge only if it is considered in its function of structural correspondence with the object that is at the origin of cerebral modulation. . . . Every structural domain, in itself, is a kind of field of consciousness, of consciousness-being. It appears as a material body only if it is known (in a knowledge-correspondence) by another being. . . . This realism of structure-consciousness clarifies the problem of the brain and of consciousness. Consciousness is not produced by the functioning of the brain, as the materialists believe. Consciousness is primary, and the brain—or a certain stage of its connections—is only consciousness, appearing as object, as body, to another consciousness."[17]

What we learn from this specification is that Ruyer's conception of primary consciousness differs equally from neuroscience and from phenomenology, both of which focus on phenomena of secondary consciousness. Thus, despite the vast difference between a materialist model of brain function (e.g., a brain image) and a phenomenological account of perception, both involve "knowledge-correspondence," the representation of a primary consciousness *to another consciousness.* As two distinct, though ultimately compatible, representations, materialist neuroscience and phenomenology both depend on a more fundamental philosophical account capable of explicating the being—the absolute form or domain—that is at issue in their representations.

ABSOLUTE SURVEY AND TRANSSPATIAL MNEMIC THEMES

The absolute primacy Ruyer grants to primary consciousness culminates in his conclusion, in the crucial chapter 9 of *Neofinalism,* that there is in fact only one mode of consciousness. Repudiating any suggestion that primary consciousness differs from secondary consciousness on account of its vagueness or lack of specificity, Ruyer instead postulates that primary consciousness cannot, like a visual sensation in secondary consciousness, be "myopic for itself." "Its field of consciousness," he notes, "will only be its own organic form, which is in principle the entire universe for it. . . . In other words, there is at bottom only a single mode of consciousness: primary consciousness, form in itself of every organism and at one with life. The secondary, sensory consciousness is the primary consciousness of cerebral areas."[18]

Ruyer's unitary conception of consciousness and his account of "cerebral modulation" allow us to position secondary consciousness as an alternative to standard phenomenological accounts of knowledge. Whereas the phenomenological understanding of perception encompasses what, on Ruyer's account, can only be categorized as a phenomenon, that is, an appearance of consciousness to *another* consciousness, Ruyer's own concept of secondary consciousness designates a derivative mode of experience (hence the qualifier "secondary") *of primary consciousness itself,* a mode in which primary consciousness experiences itself (not as absolute being or self-relation), but insofar as it is modulated through its contact with external influences. In his explication of this alternative, Colonna pinpoints the specificity of Ruyer's account: "in the case of secondary consciousness, the taking into account of the exterior environment is carried out not through 'opening' or 'transcendence'—since it remains a mystery how consciousness could exit from itself in order to go in search of things outside—but by modulation of the cerebral surface. No more than any other living cell, neurons do not have the property of representation, of auto-illumination, that characterizes secondary consciousness. But their specific character is to be modulable by exterior influences."[19] The representations of phenomenological perception (as well as of brain imaging) are thus, in a sense, *twice removed* from their source in primary consciousness: they are representations of a secondary mode of primary consciousness, of its own experience of the impact of the exterior environment on its primary self-relation.

On this score, Ruyer's differentiation of primary from secondary consciousness would seem to have a strong affinity with the account of autopoiesis developed by Chilean biologists Humberto Maturana and Francisco Varela in the late 1960s and 1970s.[20] The notion of modulation parallels Maturana and Varela's concept of perturbation, and their account of organizational closure effectively positions the organism (or autopoetic system) as some kind of absolute form. Indeed, this comparison serves to highlight one of the important sources for Ruyer's reconceptualization of form that has not been addressed thus far: the postwar development of cybernetics central to Ruyer's 1954 book *La Cybernétique et l'origine de l'information* (Cybernetics and the origin of information). Just as Maturana and Varela's formulation of autopoiesis is a conceptual alternative to cybernetic models of input and output, Ruyer's criticism of mechanistic cybernetics serves to introduce an alternate conception of information as the product of processes of true form or absolute survey.[21]

Yet these parallels only go so far in the sense that they fail to address the philosophical specificity of Ruyer's critique of mechanistic science. Unlike autopoiesis, and other models from second-order cybernetics, which ultimately accept the position of science to explain system complexity through emergence, Ruyer's project remains premised (as we have seen) on the incapacity of any crowd *(amas)* to produce unity. That is why Ruyer's philosophical account of primary consciousness and absolute form must be said to differ in kind from autopoiesis and other *scientific* models of system complexity: from Ruyer's perspective, not only is a philosophical (indeed, metaphysical) account of absolute subjective form necessary to explain the fact of absolute form but it must be developed, in contradistinction to empirical science, on the basis of a transspatial and transtemporal reality.

Ruyer's project in *Neofinalism* is precisely to develop such an account of the genesis of absolute form on the basis of a transspatial and transtemporal domain. To do so, he turns to the scientific field of embryology, specifically to a philosophical accounting of the operation of embryogenesis in relation to transspatial and transtemporal "mnemic themes." The notion of mnemic theme, as Ruyer explains it in his autobiographical essay "Raymond Ruyer par lui-même," helps to specify further how his account of modulation differs from the autopoietic notion of perturbation: "At every instant [of embryological development], the current situation of the developing organism plays the role of a constellation calling [*une constellation d'appel*] for mnemic themes that take over from an embryonic site and *pass* into time by modifying its structure."[22] In contrast to the autopoietic vision, where structural change is a reaction to perturbation by the environment, in Ruyer's account of the genesis of absolute form, structural modification is a result of the influence of mnemic themes that operate through development in space and time.

The central role Ruyer accords transspatial and transtemporal themes explains why he takes recourse to an out-of-favor, if perennial, tradition in biology, namely, finalism. Given Ruyer's antipathy to mechanistic science, it is crucial that we appreciate the twist that he brings to this branch of biological science; for in calling his finalism a "neofinalism," Ruyer does not simply mean to emphasize its rejuvenation of traditional commitments. Rather, he seeks to highlight its fundamental relocalization of the operation of finalism—from individual organisms and biological entities to the entire system of biological operationality. Again, Ruyer makes as much clear in a stunning and beautiful passage from his autobiographical essay: "In the invention or creation of forms—as much in human technics as in biological evolution—norms or directing essences play the same role as mnemic themes in this kind of *false invention* that is memory. . . . I knew already that the property of fusion and doubling of lineages of individuality made it possible to conceive of a generalized evolutionism, a generalized theory of 'descendence with modification,' in which the human is in temporal continuity, not only with an animal ancestor, but with a single-celled organism, a virus, a molecule, in which all the beings can be considered, on the plane of spacetime, as avatars of a common primordial being, just as they can participate, in the transspatial order, not only in a kind of specific 'I' or 'memory,' but, through invention, in a universal 'I.' In the end, this conception of consciousness-activities in participation implies a new finalism, not only at the level of individual activities, but in the system that is itself comprised of all the individual activities. It is necessary, finally, to postulate a region beyond the transspatial domain; in its dimension of 'nature,' obedient to non-mechanical and non-geometric laws that nonetheless remain natural, this region can only be called theologic, since it was the source of all individualized activities, of all forms and all laws."[23]

THE FUTURE POTENTIAL OF RUYER'S THOUGHT

Let me close this introductory exposé of Ruyer's system by reflecting on some lineages that connect his thought to contemporary issues in philosophy and cultural theory. There are of course direct lineages, with Deleuze most famously, where Ruyer's master concept of the "absolute survey" is marshalled to describe the infinite time of thinking; as a figure of the brain that does not spatiotemporally decompose it, as does contemporary neuroscience, the absolute survey provides a nontopological, transspatial, and transtemporal model of the brain.[24] There are also some significant engagements by phenomenologists, most notably Merleau-Ponty, who explicates Ruyer's neofinalism in the context of his own interrogation of embryology in the lectures on nature from 1960 to 1961.[25] And of course, there are critical engagements by later phenomenologists, mostly in the French tradition, where Ruyer's thought provides a pathway to linking intentionality and subjectivity to its biological foundations at the same time as it gets interrogated for its lack of a theory of perception (Renaud Barbaras) or for its exclusive identification of the absolute domain with consciousness (Roger Chambon).[26] These efforts to expand the scope of Ruyer's crucial redirection and generalization of finalism provide exciting potentials for future theoretical work.

Yet it is to more indirect—and still to be developed—lineages that I would turn in seeking to forecast the extent of Ruyer's potential impact on contemporary thought. Brian Massumi, Paul Bains, and Liz Grosz, to name just a few exemplary cases, have all taken up Ruyer's work to extend the biological foundations of subjective experience, and while all three work in the Deleuzian lineage, they all move beyond Deleuze's (or rather Deleuze–Guattari's) explicit focus on the figure of the absolute survey to address ways in which Ruyer's general philosophical project might shake up our understanding of the nature–culture divide, the status of the subject, and the role of evolutionary thought in contemporary theory.[27] These engagements with Ruyer's philosophy open new avenues for thought that will certainly gain traction once his work becomes more widely available in English.

Even more far afield, other, still-to-be-developed lineages might prove decisive for whatever influence Ruyer's thought might come to have on the direction of contemporary critical theory. There is, for example, a significant overlap with Alfred North Whitehead's philosophy of organism, which has enjoyed a critical renaissance in recent years. Although this renaissance has focused largely on how Whitehead's process thought can open new vistas in our understanding of perception, in our conception of critical theory's relationship with practices in the sciences, and in our appreciation for the environment's impact on experience, Whitehead's philosophy (as I have argued elsewhere[28]) focuses so intensely on the ontology of process that it manages to neglect how the achievement of the speculative genesis of actualities can provide the basis for a robust account of experiential entities (what Whitehead calls "societies") at all scales of being. With his unequivocal privileging of primary consciousness and his categorical differentiation of form and phenomenon, Ruyer's philosophy furnishes a mechanism to account for experiential compositions that are singular precisely because they are neither "crowd" formations nor mere physicobiological emergences but rather "forms" in their own right. Whether and how this specific critical lineage, and many more that remain equally potential, will come to bear fruit can only remain a matter for us, the lucky readers of Alyosha Edlebi's excellent English translation of Raymond Ruyer's masterwork, to decide.

1

The Axiological Cogito

Today the problem of "God's existence," along with the problem of "God's attributes," is obsolete. In any case, the form of this problem suffers from an unfortunate contamination of philosophy by religion (a primitive religion even). As with countless religious or semireligious notions, the spontaneous question today is no longer "is it true?" but "what does it mean?" The substitution of a problem of sense [*sens*][1] for the problem of existence is typical. In fact, genuine atheism is defined much less by the lack of belief in a being called God than by the lack of belief in the existence of sense in the universe.

We clearly benefit from posing a problem of sense instead of a problem of existence. Even those who are inclined to answer negatively have at least the pleasant impression of knowing what they deny, whereas in the case of the traditional questions, "the fight between [the theist and the atheist] is as to whether God shall be called 'God' or shall have some other name."[2]

The parallelism between the problem of God and the problem of Sense can also be observed between the types of arguments. The a priori or ontological argument becomes, in the order of Sense, the axiological cogito.

Just as the ontological argument claims to prove that it is contradictory to deny God's existence, so the axiological cogito tries to show that it is contradictory to completely deny finality and sense in general. But whereas the ontological argument, in many of its classical forms, seems to be a deplorable sophism, the axiological "cogito" is perfectly irrefutable.

It is absolutely clear that at least one being in the universe "offers" a sense: man. Not man in general, but each man, each "I," when he is the subject who speaks and acts. Each "I" quite easily finds the others "absurd" and is inclined to welcome the numerous and ingenious systems that consider humans as puppets driven by pure causes. But only a few speculative sophists can pretend not to exclude their "speaking person" from the domain of validity of such systems. It is quite clear that to affirm in general that every act is a pure effect of causes and has neither end nor sense is to utter an absurdity, exactly like some lunatics who say "I am dead" or "I do not exist." For the one who affirms something affirms it to be true and admits therefore that he sought the true, which is fundamentally incompatible with being driven by pure causes. Let us give a few examples.

a. A strict dogmatic *behaviorist,* who does not turn behaviorism into a simple provisional method, affirms that the behavior of human beings, including himself, can always be described in terms of responses to stimuli and that the connection between stimulus and response, however complicated it may be due to intermediary mechanisms, always has the character of a causal chain and is realized step by step [*de proche en proche*] in strict conformity with a mechanical causality. But if, according to our hypothesis, the behaviorist's spoken or written statements are mere responses to stimuli, how and by what right can he believe that he is more correct than his opponents, the "psychologists of magic and superstition"? His responses, like the reddening of litmus paper, are real facts. But "fact" is not synonymous with "true proposition," and the responses of his opponents are facts as much as his own. Why would truth-value attach to some and not to others? Let us imagine that to a behaviorist defending his system, someone impolitely replies, "What you are saying is meaningless." The behaviorist will probably be offended; and yet the interlocutor has simply reiterated the very thesis of the one he is attacking. If, by contrast, an admirer exclaims, "You are right, how true!" his endorsement will be a refutation: a pure effect can be neither right nor wrong. Purely "causalist" doctrines can be equally refuted by approval and by criticism, whereas the doctrine of "sense" can be equally confirmed by rejection and by endorsement.

b. Köhler pokes fun at the behaviorist thesis, as well as at the associationist thesis in general, and ironically regrets that it is not true. In fact, he says, "I have promised a New York publisher to send him the manuscript of this work in a few months. I have to write it in English even though my mother tongue is German. What a pity that I cannot let my responses to stimuli take their course."[3] In fact, he experiences an unpleasant feeling before the difficulties of his task, "an obscure pressure . . . which tends to develop into a feeling of being hunted."[4]

His thesis, the well-known thesis of *Gestaltpsychologie,* is that an act has a dynamic, nonmechanical nature and that the psychological tension of the task, of the end to be achieved, corresponds to a dynamic tension on the physiological plane. The sense, the order of the acts in space and time, "is a faithful representation of a corresponding concrete order in the underlying dynamic context."

The "Gestaltist" thesis sounds better than the behaviorist or mechanistic one; it seems to do greater justice to the reality of tension, to the reality of the guided effort. But philosophically, it is no better. If the writing of his manuscript simply corresponds to the establishment of an equilibrium in his "underlying physiological context," we do not see why Köhler should be more concerned than if he had let his responses to stimuli take their course. Above all, why would his manuscript have the least philosophical value, the least truth-value? Once he has finished writing, the author will simply reach a more pleasant state of relaxation, free of the "obscure" inner pressure. He can, of course, answer that this relaxation will not be achieved unless the task is not only brought to an end but is "successfully carried out" and is fruitful in his eyes. If this is true, then we are obviously no longer dealing with a pure dynamism. Ultimately, if the task is successful, there will be a coincidence not with a state of equilibrium but with an ideal; and the prior activity will have had a sense, not only as a vector in physics, but as a conscious intention.

c. To cut matters short, let us introduce at once several other representatives of the general thesis that claims to explain human activity by impulses a tergo and not by an effort to conform to norms: a mechanistic biologist; an old-fashioned psychiatrist who recognizes only physiological disturbances; a Freudian psychoanalyst; an Adlerian; a Marxist sociologist; and a disciple of Pareto. Let us imagine that they are all listening to a man expressing, with abnormal brio, his political opinions to a friend:

THE PSYCHIATRIST: This man is having a hypomaniacal fit.

THE FREUDIAN: This defiance of authority betrays an infantile hatred of the father.

THE ADLERIAN: For what inferiority is he trying to compensate?

PARETO'S DISCIPLE: What are the "residues" under the verbal "derivations"?

THE MARXIST: What class interest does he obey? He's a bourgeois intellectual having a pseudo-democratic crisis.

RABAUD'S DISCIPLE: This is the simple effect of an upset metabolism, perhaps a calcium deficiency.

All these interpretations are valuable, provided they only claim to define elements that disturb a fundamentally autonomous activity, an activity that finds its law in the fidelity to an order of truth or ideal validity. At bottom, the friend who really listens and tries to understand and to judge is right to look for reasons behind these outbursts. But if the expert interpretations claim to be self-sufficient and to do away with the simple question of knowing whether the speaker is right, whether the outbursts have a sense, they become absurd. In the first place, they contradict one another. X's political opinions cannot be explained at one and the same time by his physiology, his infantile complexes, his libido and his class interests. Of course, the learned diagnosticians could reach an amicable compromise and create a parallelogram of forces whose outcome would be the patient's behavior. But the curious quarrels between psychoanalysts and Marxists, for example, show that such a compromise is hardly possible and that they contradict one another because each claims to explain everything.

If the materialist or the psychoanalyst looked for the true causes of human actions with a fierce and heroic care for truth, these doctrines would immediately regain their validity; but they would have to disavow, at the same time, their hegemonic claims. The doctrines are nothing more than contributions to truth. Their advocates can then say, like Max Weber, "Truth is the only thing that is true."

Lequier, as is well known, discovered the axiological form of the "cogito" from a different or apparently different angle, that of freedom: "I seek a first truth, therefore I am free. Freedom is the first truth I sought, since the search for knowledge implies freedom, the positive condition of the search."[5] The structure of the argument is the same. At bottom, even the content is identical, because the freedom thus discovered is correlated to the end and to sense. Freedom consists in working toward an end according to a norm (in this case, the rule of the search for the true). It is synonymous with finalist activity and not with "free will," pure "spontaneity," "unpredictability," or "absolute existential freedom." It is not incompatible with every motivation but only with a causality a tergo.

Renouvier systematized Lequier's argument by elaborating a statement of his—"Two hypotheses: freedom or necessity. To choose between one and the other, with one or with the other"—and by complicating it with the notion of a morally superior choice, in accordance with practical reason.[6] We shall leave these complications aside and shall borrow from Renouvier the form of the *double dilemma*. He writes, "Lequier has shown that the choice required by the alternative 'necessity or freedom,' if it is considered in the determination of the philosopher's consciousness, depends on the same alternative considered in re or with respect to the external truth of the matter." There are only four possible hypotheses:

1. Determined, I affirm my determinism.
2. Free, I affirm my determinism.
3. Determined, I affirm my freedom.
4. Free, I affirm my freedom.

Hypotheses 1 and 3 have to be eliminated, because they lack any possible truth-value. In this case, my affirmation has just the appearance of an assertion: it is the effect of a pure cause a tergo. Hypotheses 2 and 4 remain. In both cases, my affirmation has a sense and deserves to be taken into account. Yet, in the same way that Cartesian doubt is identical to the certainty of existing, if I affirm determinism as a truth, this affirmation amounts to affirming that I sought the truth. One can only seek freely. The affirmation and the negation of freedom amount to the same thing; the negation of freedom—in words or in my philosophical consciousness—amounts to affirming it in re. It is clear that the whole force of the double dilemma is borrowed from Lequier's argument.

The complexity of the Renouvierist form is not without its danger. It risks making an impregnable argument look like a sophism. Nothing is easier than to caricature the argument for the inattentive reader. Suppose I have to prove my infallibility and not my freedom. I can then say

1. Fallible (in fact), I affirm my fallibility.
2. Infallible (in fact), I affirm my fallibility.
3. Fallible (in fact), I affirm my infallibility.
4. Infallible (in fact), I affirm my infallibility.

Statements 1 and 3 should be eliminated: because I am fallible in fact, what I say does not count. Statements 2 and 4 remain. But 2 is contradictory. Therefore 4 remains. QED. Obviously this is just a caricature. Unlike determinism, fallibility does not absolutely disqualify my assertions. The contradiction in 2 shows that the affirmation of infallibility implies a logical contradiction, and that is precisely what has to be eliminated.

What gives the double dilemma its sophistic aspect, even in its legitimate applications, is that the statements of the alternative in re (or as Renouvier says, "with respect to the external truth of the matter") are written or spoken philosophical statements, which imply hypothetical stances and not real facts. "The hypothesis that *x* is a fact" is not equivalent to "*x* (as a given fact)." When I say in the double dilemma, "Determined, I affirm . . ." or "Free, I affirm . . . ," and so on, the argument means "determined in fact" or "free in fact." But, because it is a matter of an argument that I am stating, the would-be fact is itself the object of a supposition, an uncertain judgment. And the proof is that, in the end, I will reject the would-be fact of determinism. The formulation of the hypothesis about the fact should thus explicitly form a first layer, which is more fundamental than the alternative in re:

1	2	3
I suppose that	determined in fact	I affirm determinism.
I suppose that	free in fact	I affirm determinism . . . , etc.

Figure 1.

So we can clearly see that the would-be facts are not facts. The double dilemma finds itself in the unfortunate situation of the classical ontological argument, which presupposes the idea of a perfect Being before inferring that perfection implies existence. It thus presupposes existence and does not prove it, because it presupposes the idea of the perfect, which is presumed to contain existence in re:

1	2	3
I suppose	the idea of the perfect	and this idea implies the existence in re of the perfect

Figure 2.

The "fact," the truth in re, is here in case 3 and not in case 2, as in Renouvier's argument. But, in both cases, in Renouvier's argument and in the ontological argument, the "I suppose" prevents us from taking the fact seriously.

If Renouvier's *double dilemma* is nevertheless valid, this is because the fundamental supposition itself (layer 1), regardless of its content, is already the express manifestation of freedom. To say "I suppose" is already to be free; it also shows that one seeks the true and knows in advance that it exists. So much complication comes down to the simple form of Lequier's argument. Every assertion, coming after a search, whatever its content may be, implies the primacy of the true, of freedom, of "sense," and of existence as senseful activity. Lequier's argument and the Cartesian "cogito" are identical.[7] They are valid only in their axiological scope.

With careful precautions, the double dilemma can be conserved as a kind of sensitive scale or an assay balance for testing equivalent concepts. It immediately proves that there is a sense in human activity and that a totalitarian philosophy of the absurd is absurd:

1. *Being a pure set of processes, I affirm that my activity is senseless.*
2. *Pursuing senseful ends, I affirm the absurd nature of my activity.*
3. *Being a pure set of processes, I affirm that my activity has a sense.*
4. *Pursuing senseful ends, I affirm that my activity has a sense.*

Assertions 1 and 3 eliminate themselves. The fact that assertion 2 is an assertion completely undermines it. So assertion 4 remains.

2

Description of Finalist Activity

The striking parallelism between the different possible contents that can be given to the "cogito," or to the most complicated form used by Renouvier, proves their equivalence. In any case, it proves that existence, freedom, signifying or finalist activity, evaluation, and work according to a norm are intimately connected. Common sense and language implicitly recognize these intimate bonds: "What do you want to do?" is synonymous with "What is the sense of your acts?" and implies, at the same time, that we are addressing a real being to whom we say "you" and not a machine made of bits and scraps, that is, that we are addressing a free being who has a will and makes an effort. The question also indicates that we will judge the value of the activity of the one we are addressing and that he will be responsible for this value.

The intimate connection between these various notions forbids their separation in analysis; it also allows us to clarify the true sense of some of them. Not every freedom, existence, or act can serve as the content of an enlarged "cogito." These three notions may be taken in a false or overly broad sense, which would render them useless.

a. *Freedom.* If the word *freedom* is taken in the sense of "freedom of indifference" or "pure spontaneity," if it is taken in the Bergsonian sense, which would imply total unpredictability in both cases, we cannot, as we have already stressed, prove this kind of "freedom" by the Lequier–Renouvier argument.

The freedom at issue here is the freedom to accomplish a task that may be judged successful or not. It is not indeterminate in the purely negative sense of the term. It is the freedom to "succeed," to give a sense to my action, instead of the freedom to elude determinism, which interests me in the problem of freedom. If I dread making a mistake, if I feel its cruel possibility, I am free. And it matters little whether my action is predictable: "Suppose that I have hit on a piece of mathematical research which promises interesting results. The assurance that I most desire is that the result which I write down at the end shall be the work of a mind which respects truth and logic, not the work of a hand which respects Maxwell's equations and the conservation of energy. In this case I am by no means anxious to stress the fact (if it is a fact) that the operations of the mind are unpredictable. Indeed, I often prefer to use a multiplying machine whose results are less unpredictable than those of my own mental arithmetic. But the truth of the result 7 × 11 = 77 lies in its character as a possible mental operation and not in the fact that it is turned out automatically by a special combination of cog-wheels."[1] The case of calculators, in which the norm of calculation becomes an assemblage of material organs, provides an invaluable guiding thread for understanding the very nature of free and finalist activity. This activity essentially consists in improvising and establishing cerebral or physical connections that allow the incarnation of the *good* sought-after results in the physical order. The example of calculators also helps us understand the connections between the order of signifying activity and the order of determinism and step-by-step causality.[2] Here the machine borrows its sense from the man who built it in view of a specific end. Just like the results it yields, this machine can be considered good or bad. It is the organ of a free center of activity.

b. *Existence.* Similarly "existence," on which the Cartesian "cogito" depends, is not any existence whatever but uniquely the existence of a center of signifying acts (defined according to what W. Stern terms the "subjective axiological a priori").

Substance-existence, whether the substance is described as "mental," "spiritual," or otherwise, eludes the probative force of the axiological "cogito." A mind manifests itself only through its cognitive activity, that is, its signifying and evaluating activity; and outside this activity, the mind's hypothetical existence as a pure substance is no more within our purview than "pure spontaneity." A "free substance" or a "senseful substance": such expressions are probably just as absurd as the expression "intelligent square."

Last, the absolute existence of the existentialists, who affirm that existence precedes sense just as freedom precedes the values, senses, and ends it "grounds," also falls outside the purview of the "cogito" or of Renouvierist reasoning.

c. *Work-activity.* By the same token, the activity we have in mind should be taken in its proper sense of work-activity. It is different from a pure functioning [*fonctionnement*] insofar as it requires the invention of means. It is the accomplishment of a task that can be considered successful or not, according to a criterion and to norms independent of the agent's whim. Looking for the quotient of a division or the premises of a syllogism whose conclusion we are given, trying to find the best arrangement of furniture in a room, trying to assemble the organs of a machine—each of these represents an authentic activity and an authentic effort, because success cannot be described arbitrarily. A truly "gratuitous" act is not an act at all. In reality, like "absolute existence" or "pure freedom," the "gratuitous act" of novelists is always obliquely monitored by the author (or by the interposed character) for its aesthetic effect or its political sense; and it becomes meaningful to the extent that this effect is sought. A poet or a painter does not need to be a surrealist to understand the advantage of exploiting dreams or psychological accidents; but because the painter or the poet seeks an aesthetic effect in this accident, he becomes active once again merely by the decision he took to remain passive before his dream in order to convey the impression of a dream. It is the same for the decorator who makes use of a kaleidoscope or the photographer who selects the frame and the layout of his shot. Our age's typical claim of dispensing with norms and values independent of the will is more ostensible than real. Humans are not so much forced to be free as to be endowed with sense; they are free only inasmuch as they are so endowed and act sensefully. Sense and end attach to all my acts, better than glue to the hand that tries to get rid of it. An antifinalist wants to prove that he is right, just as a supporter of the "philosophy of the absurd" is convinced that he espoused the only reasonable attitude.

On the other hand, there is a danger in restricting the sense of the word *work* to "industrial or agricultural work." Modern philosophy often commits the error for which Greek philosophers are justifiably reproached, for they deemed the slave's labor to be socially inferior and were thus led to overestimate pure speculation. Contemporary humans are immersed in a particular technical and economic civilization; they perpetually wield economic tools. Philosophy is tempted either to curse instrumentality (cf. Bergson, Scheler, Heidegger, Jaspers, Gabriel Marcel, etc.) or, conversely, to restrict the signification of "work" to industrial work. Metaphysically, the notion of work-activity is quite fundamental. Work-activity is tied to existence and freedom. Every definition of existence and of freedom is hollow if it does not implicitly postulate this relation: freedom = existence = work. To eliminate one of these three terms is to discard the others. A being is an authentic being, that is, a free being, only to the extent that it makes an effort. By definition, every actual existent actualizes, that is, works. A being that ceases to work, that no longer accomplishes any act, that lets itself drift, is quite obviously no longer free. "Free" is an attribute that cannot apply directly to a substance-being; it applies only to an act or to an acting being. *Free substance* is a contradiction in terms; *free act* is a pleonasm.

Work-activity, freedom, and proper existence are inseparable from three other notions, and they bring these equally inseparable notions to light:

d. *Finality.* A work-activity is defined by an end, because an activity is by definition not a simple succession of causes and effects driving one another. It has a sense that is not merely vectorial. A sense-less (i.e., directionless) trajectory is poorly oriented toward an end, a trajectory that leads to nothing. *Sense* and *end* are nearly interchangeable words. But the terms *end* and *finality* are linguistically more specialized. Moreover, Whitehead's argument, "It is absurd to have as one's end the proof that there is no finality," does not appear as decisive as the one from which we started, "It is absurd to claim and to signify that nothing has a sense," even though the two arguments are naturally equivalent. In everyday language, the end of an action designates its goal more often than its sense. While the sense of an action envelops the totality of this action, just as the sense of a pronounced sentence envelops or "surveys" [*survoler*] the words' temporal succession, its end designates a final state that takes place in space and time, like the phases of the action that tend toward it.

Finality in this narrow sense can easily shock the philosophical mind, because it appears to imply a logically contradictory causality of the future. If, to remedy this contradiction, we admit that the material future end is present in the form of an actual idea, we no longer risk a logical contradiction but rather the reduction of finality to pure causality, in the form of a "causality of the idea"—of an idea that is conceived, not as a general nonlocalizable theme of action, but as a simple ring in a causal succession.

Last, by a more advanced specialization, at least in everyday language, finality almost always acquires a utilitarian aspect. When we speak of the goal of an action, the word *goal* evokes the idea of a material, economic value more often than an aesthetic, moral, religious, juridical, or pedagogic value or ideal. To travel for pleasure often seems synonymous with "to travel aimlessly." Some measure of this ordinary sense always risks slipping into the philosophical use of the term. The Bergsonian critique of finality was influenced by the utilitarian aspect of this concept.

When, instead of the actual finality of an ongoing activity, we consider the "fossil" finality of an industrial machine, a mechanism, or an organic assemblage, there is an even greater temptation to pose the question of finality in the form "What purpose does it serve?" And when, in the case of the living organism as a whole, we can no longer ask "What purpose does it serve?" as in the case of the heart or the spleen that conserves the entire organism, we have the impression that we are exiting the domain of finality, even though we are merely exiting the domain of utilitarian finality.

The notion of "finality without end" is the by-product of this error. It is an error that rectifies another error and not a profound philosophical discovery. The notion of "finality without end" seems very subtle, but in fact, it smacks of "philistinism," for it presumes that "true" finality is utilitarian. An organism strives to conserve itself, but more profoundly, it strives to exist, that is, to actualize values in general and not simply to struggle secondarily (with the aid of subordinated and useful mechanisms) against intoxication, asphyxiation, or desiccation.

e. *Invention.* Every work-activity presupposes an effort of invention, first in the determination of the thematic end into a more particular (albeit always thematic) goal, then in the discovery of the means for attaining this goal and successfully carrying out the work. A work in the proper sense, that is, an axiological work, implicates a creation of form. It can never become a pure functioning without degrading; it is not a set of movements according to the ready-made links of a machine or according to the differences in potential of a field of forces. In the domain of physics, work-activity corresponds not to "work" (force displacing its point of application) but rather to "action" (energy multiplied by time). The "action" of classical physics was found to be a statistical phenomenon, setting in play a great number of elementary actions, each of which probably has the character of work-activity. In quantum physics, "action" is a creation of form and not a functioning. In the domain of classical physics, this character is masked by statistical effects.

f. *Last, value.* Because a genuine work can be deemed successful or unsuccessful, it obviously implies the notion of value and, correlatively, of the norm or rule that has to be followed, either to achieve success and validity or to judge the value of the work. There are as many orders of work as there are orders of value. There is theoretical, artistic, moral, juridical, political, social, pedagogic . . . work. As a consequence of this, there are as many species of the axiological "cogito": "I strive to know . . . ; I strive to achieve an artistic expression . . . ; I strive to teach . . . ; I strive to enrich myself . . . ; therefore I am."

To this description of "sense" (and of the six notions that are at one with the idea of sense), we should add a crucial corollary. Every senseful, free, valid, inventive activity is by definition opposed to the notion of a pure functioning, a pure succession of numberable causes and effects coming one after another in a well-defined spatiotemporal order, without any possible reversibility—a succession that is, by definition, incapable of surveying itself. From this point of view, the four-dimensional world of classical relativist physics, with its "universe-lines" where past and future events are in their places, represents nothing more than an infinitely flat schema, in other words, congenitally incapable of containing real, that is, active, existents. If we believe that the ordinary physical world is consistent with this schema of classical relativist physics, then we have to admit that the description of senseful activity forces us to posit an "other world," in another "dimension" (in the nonrigorous sense of this term, because it is obviously not a matter of a fifth dimension affixed to space-time): the ideal world of values and essences, to which the consciousness at work is addressed, both to aim for ends and to discover means. By definition, ends and means do not exist as such in the world of causes and effects, or at least, they cannot be encountered in numbered places along a universe-line. The sense of an activity is what this activity is *not,* in its literal unfolding. The sense of a voyage is the "end" of the voyage, in the double sense of the word *end.* So a dualist conception of two worlds, real world and ideal world, is necessary for understanding sense, finality, work, invention, and conscious existence. If the schema of classical physics is taken to the letter, it is clear that activity proper requires the positing of an ideal domain that is irreducible to the plane where causes and effects succeed one another. In this ideal domain, conscious intention can move and survey (without strict spatiotemporal localization and by exploring possibilities) the plane of causes and effects so as to influence the unfolding of means toward the ideal end. This duality of two worlds is not the last word on the question; but if we want to correctly describe and "situate" senseful activity, then the compensatory hypothesis of an ideal world is the inevitable counterpart of the fiction of a world of universe-lines or of pure causal lines.

If the numbered arrows in space-time (Figure 3) represent the gestures of a traveler who dresses, hurries to the station, buys a ticket, and boards the train, it is clear that the description of this movement, seen as a mere succession of causes and effects in space-time, has to be completed with a description of the sense and end of the traveler's activity, sense and end that "survey" the unfolding of causes and effects and organize it into a signifying whole. In other words, all the notions we have described are characterized by a *unitas multiplex,* to borrow W. Stern's expression. If the multiplicity is "realized," then unity has to be considered as "surveying" *(survolante).* If not, then the *unitas multiplex* can be expressed with the single word "form." Every activity, every conscious existence, has a form; and each product of a finalist activity presents the observer with a complex structure. In the product-structure (in contrast to the activity-form), the multiplicity immanent to the form has been "realized," as in a machine in which the pieces assembled by the engineer propel one another.

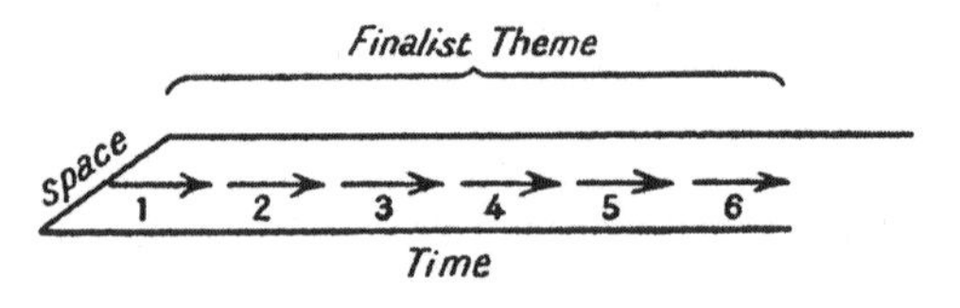

Figure 3.

We have not dwelt on *sense* itself. We consider it sufficiently illuminated by the analysis of notions that are at one with it. Every attempt to define sense, which claims to supply the "sense of sense," can only muddy the waters.[3] Let us confine ourselves to stressing that sense is not signification in the etymological sense of the term, that is, the designation of a sense by a sign. The existence of signs and significations implies—in a different sense than the existence of machines—a striking dissociation of two planes: the plane of multiplicity on which signs exist in their physical succession and the plane of the transcendent unity of the designated sense.

Suppose X is speaking to me. I grasp or spontaneously search for the sense of his words, what he means to say. But if he acts without speaking, I have exactly the same attitude: I search for what he means to do. I have again an identical attitude before an animal, assuming I do not burden myself with behaviorist hypotheses. It matters little whether the action I witness has or has not, in addition to its proper intention as an action, a signifying intention toward me. Whether I grasp it or not, the sense of an activity is inherent to it and does not depend on a witness. It is therefore a very lamentable error to define sense through the much more specific notion of signification.

Humans are so habituated to language—that is, to "signified" sense—that they easily doubt the sense of what does not speak, of what does not express itself through spoken or written words. They imagine that they give a sense to things by naming them.

3

Finalist Activity and Organic Life

Can the "I" of the human that I am, the center of senseful activities, be isolated? Can it be posited in the void, a metaphysical foundling?

Historical reasons can help explain existentialism's systematic—or, rather, nonchalant—sidelining of the problem of organic life. As F. Bollnow showed, existentialism is a radicalization of the philosophy of life, which is represented first and foremost by Dilthey.[1] *Interested particularly in the questions of philosophical method, Dilthey took "life" in a fairly vague sense, wavering between "the life of the thinking individual" and "human life in general." For him, life was the common source of heterogeneous theoretical, aesthetic, and religious activities, which can be understood as works of consciousness and not explained as things. Biology proper, whose method seems at first blush purely explanatory, did not interest him, and the material organism appeared to him more "thing" than "mind." Existentialism rectified the vagueness of this notion by accepting and intensifying Dilthey's dissociation between the life of human consciousness and the organic life studied by biology. The human* Dasein, *a much more precise notion than "human life," no longer has any conceivable connection to the human organism. Existentialism's violent and paradoxical originality stems in large part from this. It is very curious to note that there are similar historical reasons for Cartesian antivitalism and for the abrupt opposition it establishes between human thought and pure mechanism: the vitalism and the vague animism of the Renaissance barred the way to a clear and consistent philosophy. Cartesian doubt is directed precisely against these muddled doctrines.*

We have no reason to follow suit. We will not shut our eyes, for the sake of philosophical purism, to the fact that the senseful activity of humans stems from their organisms. The correct way to avert the vagueness of a philosophy of life does not seem to lie in ignoring life purely and simply or in interpreting it according to a more ambiguous dialectic. On the contrary, we have to examine how senseful activity can emerge, not from "life" in the vague sense, but from the apparently material organism, on which biology teaches us precise lessons.

The man who speaks—heard by a friend or a psychiatrist—could not speak if he were a pure mind without a larynx or a tongue. Eddington insists that his final mathematical formulas are "the work of a mind that respects truth and not of a hand that obeys Maxwell's equations," *but he needs his hand to write.* If there is a cognitive activity according to a sense and to an ideal norm, if the hand is guided in its movements by sense (thanks to cerebral links improvised by cognitive activity), the hand itself in its organic constitution and as a living organ also has a sense and had to be constituted in the first place according to a sense. The man who calculates and who, to spare himself the fatigue of mental arithmetic, prefers to resort to a calculator avails himself of what other men built in view of such an economy. Thanks to the machine's gears or its "mnemic" recorders, he does not need to keep in his mind the digits he wants to manipulate; thanks to the electric motor and the printing mechanisms, he does not need to use his hand to write. Tools and organs are interchangeable, vicarious. Both presuppose sense and finality, as much in their construction and constitution as in their use.

If it is absurd, as we have amply demonstrated, to deny sense in the human activity that seeks the true (or a political or economic output or an aesthetic effect) and leads to mathematical propositions, to calculators, to works of art, to well-adapted institutions, it is equally absurd to deny sense in the organic activity that constitutes the organs, because the organs conform to the same norms of utility and of aesthetic and technical productivity. It is even more absurd because it is through organs that man's finalist activity can build tools and other works of culture.

But it is best to consider a simple and schematizable example. We borrow it from culinary art, which is in the end also a form of culture. Dilthey could have considered it in his review of the works of the mind, and it would have allowed him to escape every temptation to separate the life of the mind and the life of the body. Let us imagine a cook at work; let us even represent him with a schema (Figure 4), which we will make extremely rough to break the association of ideas. His activity has a sense: he seeks to obtain the right result; he strives to follow the rules of the art, inventing new procedures at the risk of error. In short, the whole constellation of notions tied to senseful activity is present here, and the formula "I cook, therefore I am" is a valid form of the "cogito." The cook uses tools (casserole, spoon) that he wields with his organs (eye, hand), which are controlled by his central nervous system and, more particularly, by the cerebral cortex. If we assume that he eats the meal he has prepared, we can say that a circuit is established that goes from his hand to the utensil, then to the gastrointestinal tract. The motor of this whole circuit lies in an organic need, whose sense and end are clear, even though its modes of action are quite enigmatic. The complex modalities, the sophistications of culinary art, are introduced through the central nervous system; and this system is in a relation, not only with the rest of the organism, but with a sociohistorical culture and with a certain normative ideal, which cannot be represented on a geometric schema. If we now want to schematize digestion, the circuit will become internal: the movements and chemistry of the stomach are controlled by autonomous sympathetic and parasympathetic nervous systems, which are intimately tied to the central nervous system, especially through the hypothalamic region. The same holds for assimilation. Obviously it is futile to establish a precise border between the internal circuit and the external circuit; the latter emerges from the former, complicates and prolongs it, and it is absurd to admit sense and finality for one and to deny them to the other. The cooking of food is a predigestion in external circuit in the same way that digestion naturally continues the preparation and ingestion of the food.

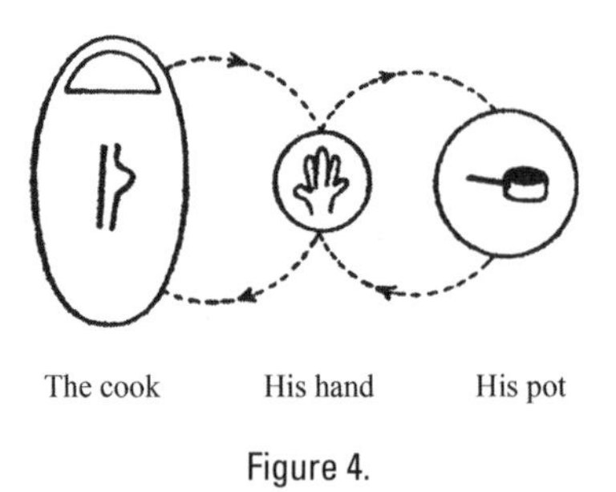

Figure 4.

Buying bicarbonate at a pharmacy is an act of the organism, just like secreting pancreatic fluid. Both of these acts have the same goal. By dint of regularly seeing X, our neighbor, go to the pharmacy or the grocery store, we no longer see in these habitual actions the biological acts that social activity covers over; or we have the impression that there is no connection between the two levels of activity. It is even more difficult to see in X, our neighbor with whom we are chatting, the embryo that his actual body continues and, in the effort he makes to speak, the sequence of efforts that this embryo had to make to constitute a larynx and a tongue for itself. A rough schema can help us to rediscover this undeniable truth.

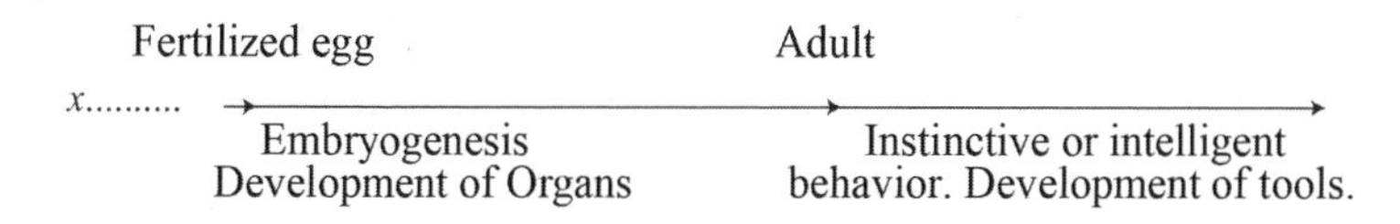

Figure 5.

For the same reason, it is impossible to recognize a finalist sense in the invention of cooking utensils and to deny it to the organs of ingestion, digestion, and assimilation. The teeth are grinding tools, the stomach a retort and an automatic mixer. The adult cook fabricates utensils or obtains them from a shop; but the adult himself, with his stomach and his brain, is the outcome of an embryonic creation, whose principle is concealed from us but whose work undeniably has a sense, because it prolongs itself externally according to this sense, through senseful technical works.

The instinctive behavior in external circuit is normally interposed between organic activity and intelligent finalist activity. Instinct, with some exceptions, does not fabricate tools and unfolds less in the "external world," in the human and industrial sense of the term, than in a biological *Umwelt* given with organism. It is moreover difficult to isolate organogenesis from the *Umwelt,* because an organ almost always has a double polarity, like a tool that has a handle and a blade, with one of the two poles directed toward the "environment" (even when this environment is internal). It is impossible not to recognize that instinctive technology prolongs organogenesis: the spider's art of weaving clearly extends the formation of his silk glands; that, more generally, animal behavior is a "regulation in external circuit": warm-blooded animals who instinctively seek heat and cold according to the needs of their bodies extend with their behavior the action of organic mechanisms that regulate the body's temperature.[2] But what is true for instinctive behavior is equally true for intelligent behavior, and in humans as in chimpanzees, the instinctive gesture is very often the germ of an elementary intelligent intuition. Only progessively is this intuition emancipated from instinct and the biological *Umwelt*. The invention of clothing, of weaving, and of the treatment of fur, and the invention of heated suits for aviators are thermic regulations in external circuit. In many cases, we can discover three corresponding levels: organogenesis, instinctive behavior, and intelligent activity. Hence the formation of organic reserves (fat, sugars), instinctive reserves (honey, various provisions), and intelligent reserves (an Eskimo's meat caches, our jams and wealth). So long as we only consider instinctive behavior, it is vaguely possible—at the price of some bad faith and a good deal of nudging in the right direction, and provided we first imagine that organogenesis itself can be explained by physicochemical causes—to argue that instinctive behavior can be explained in the same way. But if we add the intelligent human behavior to the series, the theory becomes untenable. Humans exist and act, and their activity *reveals* the true nature of organic activity. At times, human activity contradicts organic activity: humans can commit suicide; they can curse life. But the man who commits suicide uses his own organs to destroy them.

To interpret the totality of the facts, we have to climb back from intelligence to instinct and from instinct to organogenesis. Because sense and finality exist in intelligent activity, they have to exist in instinct and in the organism. The mode of this finality can and must be deemed different for each level; profound differences can exist between organic finality, instinctive finality, and intelligent finality. But it is simply impossible to admit an absolute difference of nature between these levels, to admit that a senseful behavior in external circuit could emerge from an organism that would be a pure set of physical phenomena connected by a step-by-step causality. If X's conversation is meaningful, the constitution of his larynx and his brain must have been a meaningful work; the organism is the first of meaningful works. Biology cannot be separated from comprehensive sciences. Admittedly, biology for the most part explains facts; it studies structures and mechanisms, and it approaches the organism in this way. This is not surprising, because the organism is a set of organs that resemble, despite their superior complexity, the machinery of human industry. But a sense dominates this whole arsenal, just as the minds of men dominate their entire machinery.

After a meticulous study of techniques, Leroi-Gourhan sought to bring biology and technology closer together. Technical intention and creation extend the instinctive movement by which the living being strives to "make contact."[3] *The evolution of techniques has to be expressed in biological images—diffusion and segregation, mutation and heredity: "If we are looking for technology's real family, then we need to orient ourselves toward paleontology, toward biology in the broad sense."*[4]

It is because the tool and the machine extend organic activity that they always remain subordinate to it and have no persistence of their own. Woodger notes, "A machine is *made* to realize some conscious human purpose. Its parts work together to secure *that* purpose, not to secure its own persistence. . . . Machines are subordinate to organic persistence; they are used by human organisms for the purpose of securing their own mode of persistence, or the persistence of something that may be valuable to them. A machine may in fact be regarded as a part of an organism of a peculiar kind, linked to the rest by psychological as well as biological ties."[5] We can add that the instruments and machines we use perish in principle at the same time as our bodies. Like the organism, machines that have not been consciously maintained become corpses whose "form" is only a structural appearance.

Mechanistic biology is not necessarily antifinalist. We can even claim that mechanistic biology is more naturally finalist than antifinalist. It seems antifinalist only because the theorist forgets that humans create the machine and that humans are organisms.

In the seventeenth century, minds as profoundly religious as Bossuet, Malebranche, and Nicole accepted and admired the mechanistic biology and medicine inspired by Descartes. "The ear has convenient cavities for making the voice reverberate in the same way that it reverberates in rocks and echoes. . . . The vessels have their valves turned in every direction; the bones and the muscles have their pulleys or levers."[6] *They admired all the more this marvelous art that for them, it goes without saying, presupposed an artist. If Cartesian anatomists like Dionis and Stenon were reluctant to speak of final causes before the mechanics of the body, it is not because they doubted its finality in general but because they feared the temerity of those who claim to know the use and the precise and particular end of this or that organ.*[7] *Like Paley and the finalists of the eighteenth century, Malebranche dealt with the famous theme of the clock. "It should be noted that all this happens mechanically. . . . This is why we are bound to admire the incomprehensible wisdom of the Being who has so well arranged all these forces, that it is sufficient for an object to touch the optic nerve lightly in this or that way to produce so many different movements in the heart . . . and even in the face."*[8]

It is certainly not by equating the organism to a set of machines that we will manage to elude teleology. Every explanation of organic teleology that relies on an analogy to machines amounts to explaining internal teleology by means of external teleology; but in both cases, a teleology is always at stake. The cruder the mechanism, as was Descartes's, the cruder the corresponding teleology. The more the human body resembles an automaton, the more God resembles an engineer.

It is typical that Cuénot-Andrée Tétry's thesis (the equating of the organ and the tool) is today deemed to be finalist in its inspiration. Rightly so.

4

The Contradictions of Biological Antifinalism

We can rediscover the preceding result directly, without even invoking the impossibility of separating the external circuit and the internal circuit in biopsychological activity. Just as it is contradictory to "signify" that there is no sense, or to have as our end the proof that there is no end, or to defend the truth of a thesis that, reducing everything to pure causes a tergo, employs the word "truth," in short, just as there are internal contradictions in antifinalism when it bears on conscious human activity, so there are internal contradictions in biological antifinalism, even when biological facts are considered objectively. In some sense, these objective contradictions are incarnated in the facts. They do not result, as in the first case, from the conflict between the form of an assertion and its content. To grasp the distinction, let us reflect on the two kinds of objections that can be raised against an epiphenomenalist who maintains this curious theory: consciousness is an ineffective glimmer that accompanies autonomously unfolding nervous processes.

1. We can respond with the same argument we examined in the first chapter: "Because you are a pure machine, your assertions cannot be true."

2. But we can also raise another objection, for example, "How could a being in whom consciousness is an ineffective pure accompaniment have invented anesthetics?" In this case, the contradiction is incarnated in the facts.

We shall now examine the contradictions of this second kind. They are as effective against the antifinalist thesis as the pure logical contradiction of the first kind.

a. What is so shocking to every unprejudiced mind about this consequence of epiphenomenalism: the unconscious invention of anesthetics?

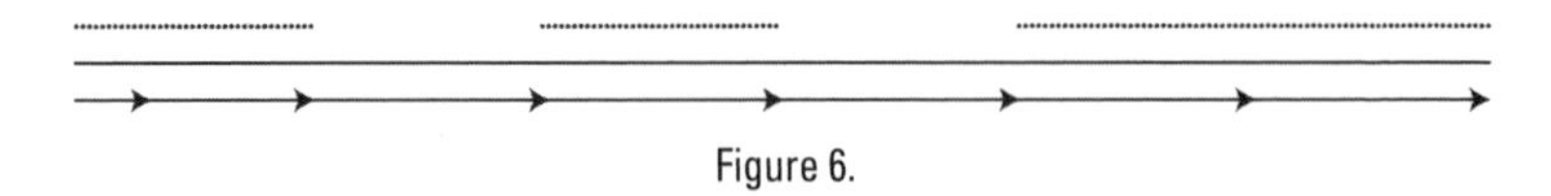

Figure 6.

Let us represent with two parallel lines the nervous process (in a continuous and solid line) and the epiphenomenal consciousness (in a discontinuous and dotted line). The continuous line thus represents a succession of causes acting on one another step by step. The dotted line cannot effectively act on the solid line; but the human invention of anesthetics presupposes that the unpleasant consciousness induced human beings to seek the means to suppress it. If, according to the hypothesis, the unpleasant consciousness is not effective, how could it be at the origin of an action? And how could a chain of pure causality manage to escape this unpleasant consciousness?

We encounter a similar situation in all cases where the living organism uses or seems to use a game of chance. And these cases are numerous. The most striking is the mechanism of recombination of genes in sexual reproduction. Biologists unanimously recognize its great significance alongside mutation in the life and evolution of species. Darlington recently showed the advantage of the various mechanisms of *crossing over,*[1] of delayed meiosis, and of the reduction in the diploid phase in metazoans, for the plasticity of the species.[2] Another mechanism is the one through which sex determination takes place. One of the two gametes is heterogametic; there are two kinds of eggs or two kinds of spermatozoids, depending on the species. Sex is thus determined by a game of heads or tails, which ensures the approximate numerical equality of the two sexes according to the laws of chance. If the organism as a whole is the result of pure causes devoid of active finality, if it is the outcome of a pure automatic sorting of fortuitous variations, then we have to interpret the fact by saying that chance fabricates a game of chance. This contradiction is just as striking as the earlier one: "ineffective consciousness seeking to suppress consciousness." In fact, in conscious activity, the use of chance is by definition always desired. It is always a voluntary renunciation of a voluntary choice. One plays heads or tails or one draws lots to avoid every bias and every implicit influence of habit, which would risk engendering an inadvertent asymmetry. By using a die, consciousness chooses not to choose; it deliberately suppresses its own action, just as when one asks for anesthesia before a surgical operation. Antifinalism in biology must then confront this curious consequence: chance establishes a game of chance to suppress the action of a finalist direction that, according to the theory, does not exist.

b. The regulation of metabolism or of the various metabolisms in the organism or, more generally, the regulation of internal chemistry presents us with another contradiction of the same type. According to the extremist antifinalist hypothesis (formulated by Rabaud, among others), it is metabolism (along with the environment) that is the sole cause of the organism's structure. This structure is not adaptive in the strict sense; it is arbitrary. Selection confines itself to eliminating the worst. Many biologists who do not share Rabaud's generalizing dogmatism adopt at bottom the same point of view; it does nothing more than coherently express the very principle of determinism in biology.

We have already demonstrated the absurdity of this thesis when it is extended, in conformity with its logic, to the psychic productions of the humans who express it. We could have also argued by relying on the intimate connections between external circuit and internal circuit. A man who intoxicates himself modifies his own internal environment by exploiting an external circuit that passes through a whole social technology. Alcohol is an "internal secretion" of human society, just as adrenalin is an individual organic secretion produced by organic technology. On the other hand, a man who has a drink of alcohol seeks, for instance, to muster courage for a difficult task. He foresees what his organic and psychological state will become after the sought-after libations and thus clearly shows that, in the inseparable organism-external-action assemblage, something completely escapes the temporal succession of various states of the internal environment, because this something uses it as a means and even eventually—as in the preceding example of anesthesia or the game of chance—to momentarily suppress its own autonomy. According to Canon, the organism proceeds in the same way when an *emergency system*[3] comes into play in emotion and a hypersecretion of adrenalin increases muscular power, arrests digestion, accelerates the movements of the heart, and so forth. But if, by hypothesis, the physicochemical metabolism is the general cause of the organic structure, then we have to conclude that it is the organic metabolism that establishes the complicated system destined to control it.

c. If, instead of dubious channels, we were to clearly see on the surface of Mars the geometric construction that demonstrates Pythagoras's theorem, the hypothesis of intelligent Martians would find few detractors. For this construction would reveal that the Martians possess a truth. The possession of a truth is obviously a whole other thing than automatic obedience to a law. No one doubts that real phenomena on Mars obey the laws of geometry or mechanics; but a proof of the possession and use of these laws by Mars's inhabitants would be absolutely "sensational." Between these two orders of facts, there is the same difference as between "following the laws of chance" and "inventing a game of chance" or between "undergoing alcoholic fermentation" and "drinking alcohol to muster courage." All of these contradictions are "isomorphic." In these cases, it would be contradictory to explain the second order through the first. (Animal camouflage, as we will see, is the most beautiful instance of this type of contradiction.)

We know that *Gestalttheorie* claims to apply its interpretations of forms not only to psychology but also to biology. Wertheimer and Koffka claim to resolve, in psychology and biology, what Koffka calls the positivism–vitalism dilemma, that is, the dilemma of "causal explanation" and "explanation by sense and value."[4] Köhler tried to show that his principles explain the regulations after lesion or after experimental excision in the course of embryogenesis.[5] The theses of "Gestaltists" have influenced biologists considerably and in every domain, from neurology (Goldstein) to embryology (Dalcq).[6] The insufficiency of the thus-generalized *Gestalttheorie* is obvious, in the presence of facts that experimental embryology has firmly established. But the facts of animal camouflage or the opposite and analogous facts of "animal publicity" (aposematic coloration of animals who want to alert predators or enemies to their identity) show not merely its insufficiency but its contradictory nature, by virtue of the principle that the same organic structure cannot be deemed at once *to obey* the laws of *Gestalttheorie* and *to use* these laws to conceal or announce itself. An organism that *uses* the laws of Gestalt cannot *be explained* by these very laws. This contradiction would be equivalent to saying that Newton's discovery of the laws of gravitation had as its sufficient cause the fact that Newton's body underwent terrestrial attraction.

In his great work Adaptive Coloration in Animals, *Hugh B. Cott follows by and large the outline of a treatise on camouflage or on public display, which seems to be inspired by the well-known laws of* Gestaltpsychologie *on the perception of forms: laws of segregation and grouping, laws of the good continuation of lines or movements, laws of pregnant forms, laws of figure and ground, laws of the figure's internal organization.*[7]

By way of example, let us consider the facts that can be classified, according to Cott's expression, under the rubric Disruptive Coloration. *For the observer, the unity of the object consists in its approximately continuous surface, surrounded by a contour that contrasts with the background. When this object is mobile, it cannot be concealed by a simple homochromy with an invariable ground, so it becomes necessary to break the contour with violently contrasted stains, some of which have a chance of merging with the ground and some of which, though very visible, constitute a different configuration from that of the camouflaged object. This is precisely what we observe in a multitude of animals from the most diverse species (in butterflies [Figure 7], fish, frogs, reptiles, mammals, or the eggs of certain birds). In many cases, the organism refines this procedure by accentuating the contrast of tones between adjacent stains (skins of reptiles [Figure 8]; frogs:* Cardioglossa gracilis; *fish:* Eques lanceolatus; *birds:* Pluvier, *etc.). All the camoufleurs know that the contrasted stains not only must disrupt the contours of the object but have to be completely detached from the object's natural elements. A priori, it seems more difficult for nature to apply this principle to an organism, whose various parts form natural anatomic and physiological wholes, than it is for the camoufleur painter to apply it to a tank, for instance (he can easily spread the stain by cutting the gun barrel, the dome, and the body of the tank). Moreover, biologists like Tylor*[8] *have claimed to show that the color marks follow an anatomical base; in this way, the patterns of snakes would be tied back to the underlying bone structure, and the same would hold for the majority of the patterns of caterpillars or birds. But in fact animal camouflage is most often as perfectly disconnected from the underlying anatomy as the most successful camouflage of an assault tank. In the fish* Heniochus macrolepidotus *or in* Dascyllus aruanus *(Figure 10), a dark, short stain runs uninterruptedly along the whole body, covering the eye and bisecting the dorsal, pelvic, and anal fins.*

In a whole host of frogs, butterflies, and grasshoppers, the camouflaging

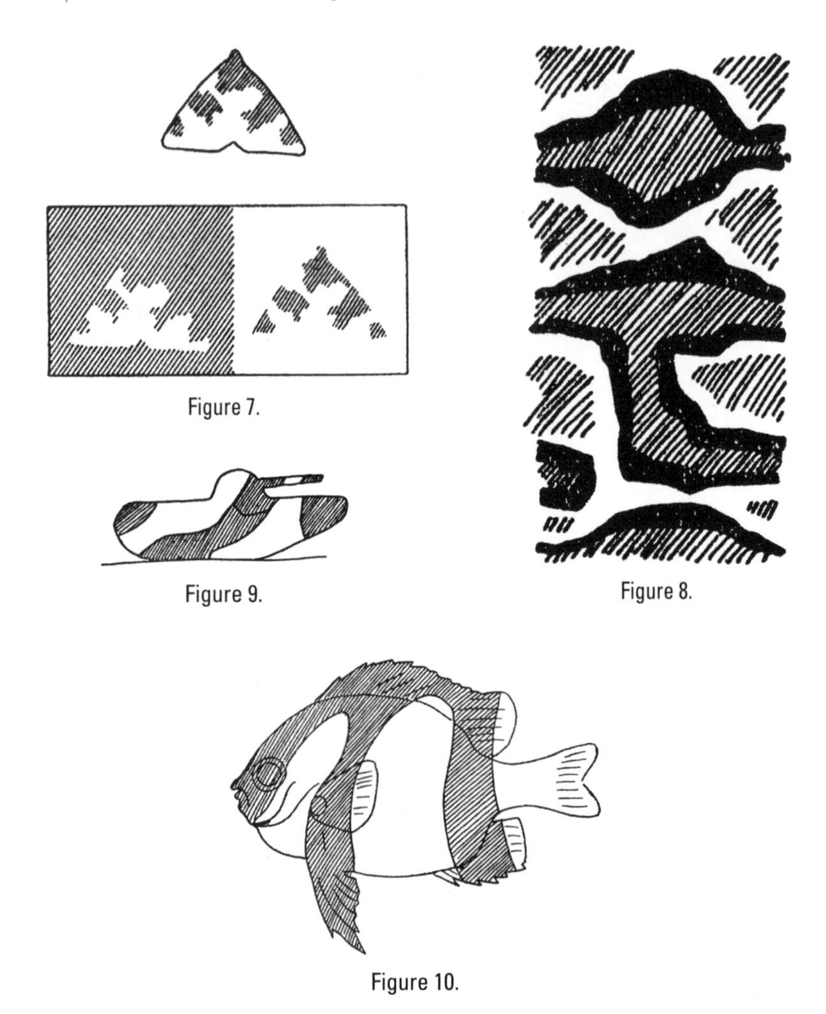

Figure 7.

Figure 9.

Figure 8.

Figure 10.

stains or bands are arranged in such a way that not only do they virtually shatter the unity of an organ or a natural part of an organ but—when the animal is immobile—they optically fuse distinct organs thanks to their harmony (cf. Edalorhina buckleyi *or* Rhacophorus fasciatus*) (Figure 11).*

The eye of vertebrates, mainly in animals without eyelids, is so constituted that it attracts attention by its circular "good form" and its dark pupil. Countless organisms have specially treated their camouflage with auxiliary stains, which have similar functions to the Gottschaldt figures we find in all the expositions of Gestaltpsychologie: *frogs (*Rana oxyrhynchus, Cardioglossa leucomystax

Figure 11.

[Figure 12]), serpents, fish, and so on. The elongated stain cuts the eye, which melts entirely into the stain by the dark part of the iris and into the rest of the body by the clear part. Among certain predator fish (Lepidosteus platystomus), *the band that camouflages the eye and that appears simple to the observer consists of a series of pigmented areas that affect seven distinct anatomic units. In other cases* (Pterois volitans), *one complex figure (Figure 13) converges on the pupil and entirely "absorbs" it.*

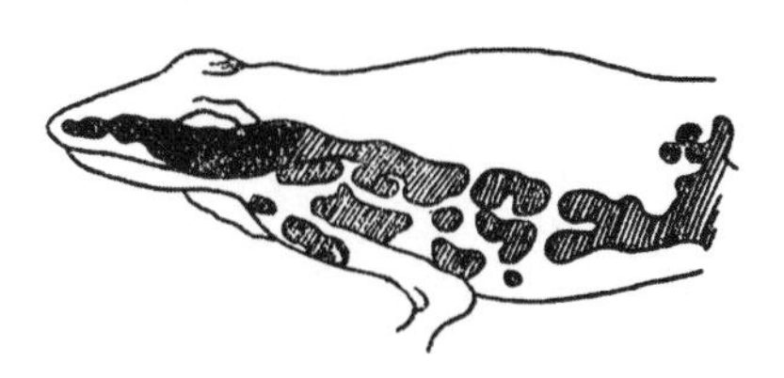

Figure 12.

Figure 13.

Among fish with camouflaged eyes, a "deflective" and extremely apparent mark that resembles an eye is sometimes situated on a nonvital region (at the end of the tail, for example) in such a way that the animal is seen illusorily inverted (Choetodon capistratus, C. plebejus, Antennarius notopthalmus). *These fish swim slowly, tail first, and, in case of danger, dash rapidly in the opposite direction. Among other fish* (Pomacanthus imperator) *(Figure 14), there is strictly speaking no false eye or any deflective mark; rather, we find an arrangement of curved lines that form, as Lewin would say, a field oriented toward the caudal region. Thecla butterflies exhibit an even more perfect optical inversion from head to tail (especially* Thecla phaleros*) (Figure 15), obtained by an accumulation of processes: pseudo-antennas, a false eye, convergent bands of wings on a false head, movement of false antennas, and immobilization of the true ones.*

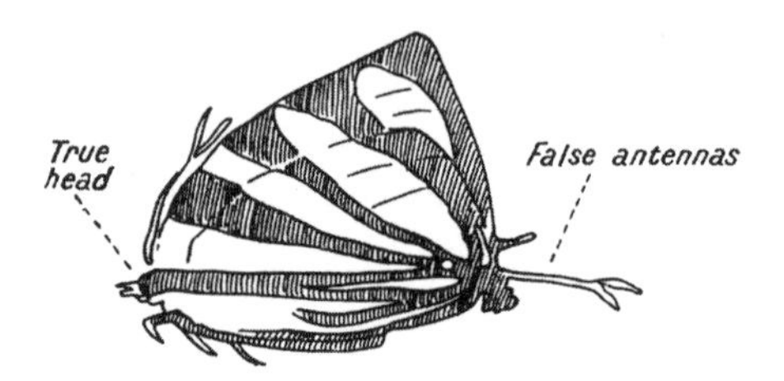

Figure 14.

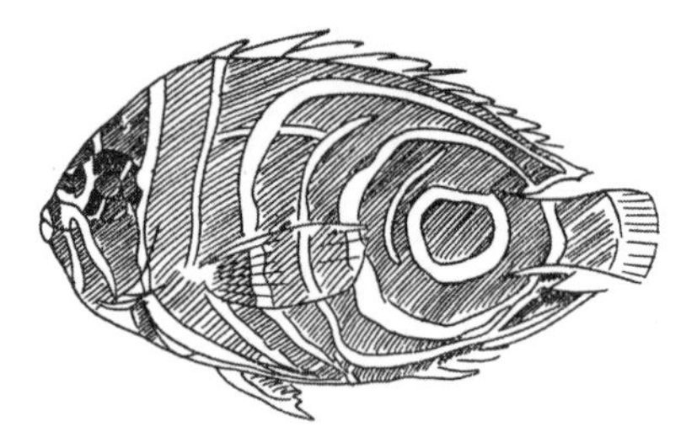

Figure 15.

This case perfectly illustrates the opposition between a possible explanation of the organism and of its behavior by "Gestaltist" principles and the fact that the organism uses *these principles to disorient its enemies. Many biologists, who are more or less steeped in* Gestalttheorie, *recently invoked a dynamic or chemical cephalocaudal gradient in organogenesis and stressed the fact that the head under formation quickly becomes a dynamic pole and the active region of the field (Child). They have even stressed that an experimental, chemical, electric, or thermic gradient can at times move the zone of cephalic organs inside the embryo (Gilchrist and Penners). But it seems that the true motor of development is very different from a simple gradient and that this gradient has an entirely occasional role, because the organism can eventually simulate an inversion. In well-known and very ingenious studies, the antifinalist D'Arcy Thompson deployed "Gestalt"-like principles (principles of least action and of maximum-minimum, gradients of hormones producing differential growth) to explain, by pure physical laws, the curious amplifications of the caudal regions in certain species of the* Diodon *genus.*[9] *But this explanation, which seems to be valid in some cases, is certainly not valid in others; and at any rate, it does not represent the whole matter. Cott (Figure 16) showed that the fish* Platax vespertilio, *which resembles a leaf by the enormous development of dorsal and anal fins, reaches the same leaf appearance as* Monocirrhus polyacanthus *(Figure 17) through the opposite approach. In* P. vespertilio, *the leaf appearance results along a line* perpendicular *to the axis of the body. Even if a hormonal gradient is the initial cause of the enormous amplification of dorsal and anal fins, this amplification was used by the organism and highlighted by dark bands and by the discoloration of the caudal fin.*

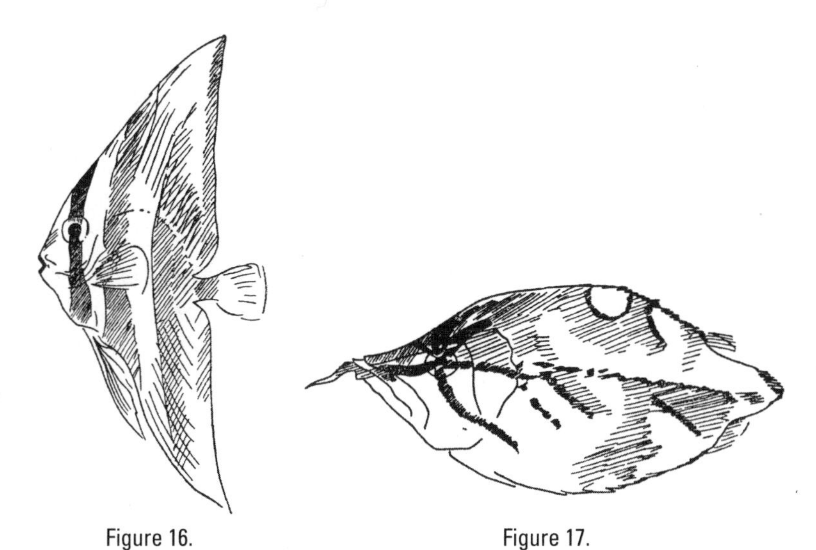

Figure 16. Figure 17.

A distinctive feature of the use *of laws by a technique, in contrast to a pure automatic* obedience *to these same laws, is that one technique uses the laws* to all ends and even to opposite ends: to construct and to destroy, to heal and to kill, for peace and for war. This distinctive feature is not lacking here: organisms use the laws of good form and other laws to camouflage themselves, but they also use them to show themselves. Organic publicity is useful in a number of cases: it attracts the attention of the opposite sex in species that are not very widespread, or it signals the identity of an animal to the predator when the species is foul smelling or dangerous. This publicity uses the same laws as camouflage, but in an opposite way: colors that contrast with the background, the arrangement of lines into warning display.

Up until recently, it was fashionable to deny all these facts as well as the associated facts of disguise and mimetism. It was only a matter, one claimed, of fanciful or biased interpretations. French biologists are generally hostile (Fabre, the finalist entomologist, considers the belief in mimetism and animal camouflage an "inanity"). As recently as 1936, A. F. Schull considered the notions of camouflage and mimetism "armchair" speculations that belong to uncritical times.[10] As J. Huxley notes, it is on the contrary these very objections that are "armchair speculations." Cott's studies were confirmed by the independent work of Stuffert (1935), Cornes (1937), Phillips (1940), and Holmes (1940). Cott showed that the facts themselves, when they are analyzed with care, respond to the classical objections: that noncamouflaged animals prosper as much as others, that camouflaged animals are not confined to habitats in which their camouflage is effective, that camouflage is merely a human impression and predator animals do not even see it, and so on.

It is important to highlight the remarkable adaptation between the kind of camouflage and the type of habitat, and the adaptation—which confirms the inseparability of instinctive behavior and organic life, of external circuit and internal circuit—between the kind of camouflage and the animal's movements and attitudes. Cott emphasized this adaptation for several of the cases cited earlier. He observed at the Regent Park aquarium the behavior of a leaf-fish (M. polyacanthus) that floats motionlessly like a dead leaf, which it perfectly resembles, maintaining its body rigid and drawing close to its prey with imperceptible movements of its nearly invisible dorsal and anal fins.[11] The Ceylon magpie described by Phillips camouflages its nest so as to make it resemble a node in a branch. Its children have the curious instinct of standing absolutely still, their beaks lifted, to imitate the end of a broken branch. Similarly, the closest resemblance can be observed between the aforementioned deflective marks (which form part of the animal organism) and the deflective marks that *certain spiders (Figure 18) instinctively create on their webs and that, made from silk and debris, have the height and appearance of the spider itself to mislead hostile birds.*[12] *On the other hand, R. Hardouin is perfectly justified in establishing a parallel between the organic fact of camouflage and of mimetism and the habits of primitive human hunters, who identify with the coveted game and disguise themselves in its plumage.*[13] *Because there are also animals that instinctively disguise themselves, like the crab* (Oxyrhynque), *which cuts the algae and affixes them to its shell, we rediscover here the three levels: organic formation, instinctive external circuit, and intelligent external circuit. This necessary parallel reinforces the argument that relies on the internal contradiction between the organic* use *of camouflage and the pure* obedience *to physical or physiological laws.*

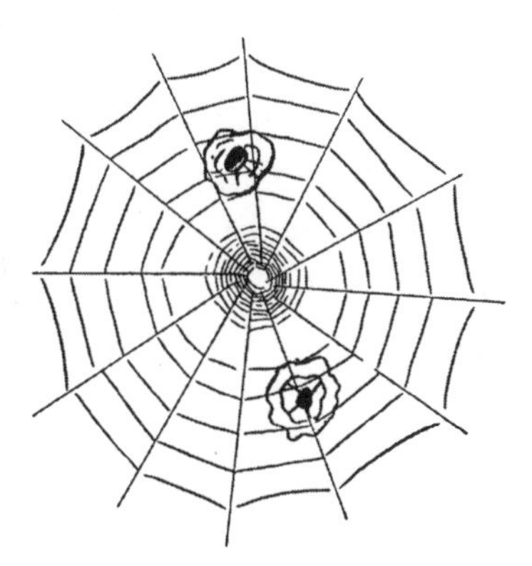

Figure 18.

5

Finalist Activity and the Nervous System

When biologists set about explaining individual evolution and ontogenesis by causes a tergo or by various mechanisms without finality, their logical errors are at times very crude. We have already highlighted the most noteworthy ones: the constant confusion of a simple trigger and an explanatory reason; the belief that a chemical can explain a structure; the verbal repudiation of the preformationist theory and the perpetual return to this same theory, which one pretends not to recognize after having disguised it.[1] The contrast between the admirable patience of observation, the ingenuity of experiments, and the extreme weakness of argumentation is such that we are tempted to suspect a psychological barrier, a conscious or subconscious decision, akin to what the psychoanalyst uncovers in his patients. Countless biologists are visibly haunted by the fear of slipping into "religious" and providentialist conceptions. The fact that the majority of finalist biologists are religious confirms the others' suspicion. We will not speak here of biologists driven by political ideologies.[2]

Nevertheless, we should not rush to accuse biologists of irrational phobias and bad faith. The phobia does exist, but it is most often rational. We see this when we examine the nature of the psychological "barrier" more closely. Many antifinalist biologists, who do not lack metaphysical sense and are even authentically religious, fear that by accepting the idea of finality in biological facts, they will be dragged not only into religious views in the broadest sense but into a naive providentialism; that they will be compelled to admit not a *Logos,* but an anthropomorphic God, inventor and fabricator. This is an infantile image and an unrealistic explanation, which double the mystery of formation and of organic invention with mythology. We will not persuade the antifinalists by assuring them that they paint a devil (God, in the present case) on the wall to frighten themselves and that today it cannot be a question of so naive a belief.

In the first place, this would not be accurate. It is true that biologists who espouse finalism as a result of religious belief do so at times in a very simplistic way. But the key to the question does not lie there. It resides in the postulate accepted by all biologists, it seems: every invention presupposes a brain or a "cerebral" consciousness. So organic finality, if it exists, has to rest on something that resembles a human consciousness, on the understanding of an anthropomorphic God. Taking his inspiration from Cuénot, A. Tétry writes, "The birth of the manufactured tool, the sign of specific human activity, is not mysterious; in general, we know the date of its creation, we know the name of the inventor. . . . The anticipated representation of the tool, i.e. of the goal or the end to be attained (final cause), conditions its production, which is then an articulated act, preceded by an idea and operating as a cause. . . . The tool is the thought-out work of the human brain."[3] But then, because "invention includes reflection, perspicacity, and intelligence," organic invention "can only be the effect of a thinking brain like the human brain." So it seems "necessary to resort to a God or to an anthropomorphic Nature . . . an option repugnant to many scientists and philosophers."[4]

Let us translate this thesis into our schema (Figure 19). In the finalist act in external circuit, the brain is an indispensible link (the cook uses his brain to make a dish), whereas in internal circuit, the stomach, for example, operates like a mixer or a furnace regulated by its own nervous centers. If we then admit organic finality by equating the stomach itself to an invented tool, it seems that a second external—"supernatural"—circuit is needed to explain this organic tool, a circuit controlled by a consciousness and even a supernatural brain. Such a duplication of humans or animals by a fabricating God—by a "gaseous Vertebrate," as Huxley said—is not very seductive to biologists.

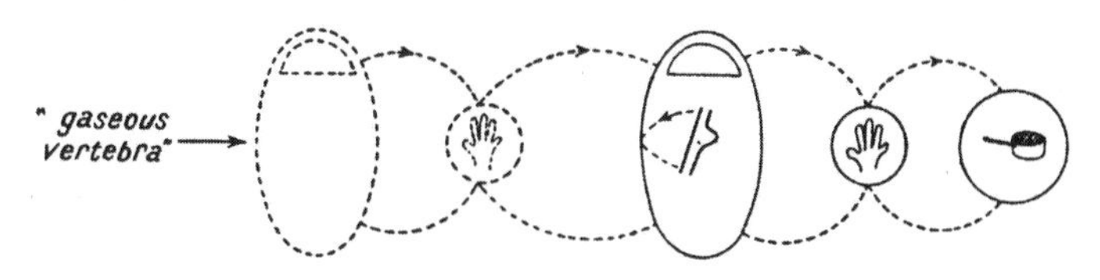

Figure 19.

But the error is clear. Finalist action in external circuit, which presupposes a good condition of the brain, is only a complication of finalist activity in internal circuit. It is thus logical to consider the brain as the instrument of this complication, *but not as the instrument of finalist action in general.* The central nervous system, extended by the eye and the hand, allows the organism to project its finalist activity into the external world; it enables it to structure and organize a vast domain, beyond its integuments and internal organs. The brain *expands* the field of organic finality; it allows finality to spill over onto the world, to discover materials within it, to construct tools and machines that are organic by their form and not by their matter. But "to transport" or "to expand" is not synonymous with "to create" or "to bring into existence." This is not a curious, paradoxical, or even personal thesis; it is not even, strictly speaking, a thesis. It is the pure and simple statement of obvious facts, of one fact above all, which no one can contest: the organism forges its nervous systems before using it. The brain is thus an "organ of transport" of finalist activity; it is evidently not its "organ" tout court.

It is extraordinary that one forgets this obvious fact as soon as one tackles the psychophysiological problem of the role of the brain. Regardless of the philosophical tendencies of the one who takes up this question, he never asks, "How should I understand this role of transport? How should I understand this transposition of organic activity into organizing activity?" He asks, "How should I understand the relations between the brain and thought, between the brain and finalist activity in general?" If he is a materialist, the brain seems to him to contain the whole secret of this finalist activity. If he is a mentalist or a spiritualist, the brain seems to him to be a mere instrument in the service of this activity. But the mentalist as much as the materialist tends to position the whole of finalist activity vis-à-vis the brain, forgetting that a primary organic activity is in any case a fact.

Before experimentally studying the central nervous system's mode of operation, we have to know exactly what we are looking for. Otherwise, an ill-posed question will only ever elicit bad responses. In interrogating experience, what we should look for is what the brain *adds* to organic finalist activity.

A physiologist who studies the gastrointestinal tract, the respiratory system, or the genital system should not and cannot forget that assimilation, respiration, and reproduction are fundamental biological functions that affect the organism as a whole before manifesting themselves secondarily in superior organisms, in large organs. The study of the nervous system, and of the brain in particular, would have to be subjected to the same rule. The essential question is to distinguish what is primary and what is secondary in organic activity, what is due to the living being, on one hand, and what is due to the particular structure of the organ, on the other (ear, lung, or brain). Prior to every concrete experiment on the role of the brain, and through a cursory examination of the facts, we see that the brain or even the nervous system cannot be deemed to have monopoly over memory, habit, invention, signifying activity in general, and finalist behavior, nor even over consciousness, conceived as the proper subjectivity of the organism.

It cannot have monopoly over memory, for the good reason that in ontogenesis, the brain is remade de novo from an egg that does not contain microstructures of the nervous system. The brain's development can be followed starting from primordia with a very simple structure (e.g., neural groove). The instrument that allows the fertilized egg to construct the dizzyingly complex architecture of the nervous system may or may not be called "organic memory." In either case, there is no question that the eventual role of the brain in psychological memory will be subordinated to what (mnemic or not) built the brain to begin with and did so without using a brain.

The brain cannot have monopoly over invention, because for better or worse, we have to recognize, even though we may shrink before the term, an organic invention of organic tools. These tools are quite similar to the tools that humans manufacture by means of their brains—similar in their form if not in their material. The human brain is certainly responsible for the fact that wooden or steel tools exist; it is surely not responsible for the existence of organ-tools, made up of living cells. The brain is indispensible for the existence of chemical or pharmaceutical factories but certainly not for the existence of the liver or the endocrine glands. The brain intervenes in the artificial camouflage of hunters or warriors, but the nervous system, combined with humoral mechanisms, intervenes in organic camouflage only for the individual adjustment of chromatophores in animals with variable homochromy.[5]

It has no monopoly over learning,[6] over the acquisition of habits, because even protozoans proved during experiments to be capable of acquiring habits, of presenting behaviors in the ordinary sense of the word and not in the behaviorist sense;[7] because there is in plants a kind of adaptive behavior very close to animal instinct; and because ontogenesis itself can be equally or better described in terms of "formative instincts" than in terms of chemical inductions.[8] If the brain is an instrument of finalist activity, then by definition the ontogenesis that constitutes this instrument has to be a finalist activity, one that can do without the brain.

Last, the brain cannot have monopoly over consciousness. This point is more delicate, not because there is any reason to hesitate over it, but because an essential and difficult distinction becomes necessary. The brain certainly has monopoly over sensory consciousness, that is, a consciousness whose "information content" is supplied by sensory organs modulated by external stimuli. Indeed, it is paradoxical to say, with Bergson and several other contemporary authors, that the area striata, for example, this cortical retina, is nothing more than a center of movement.[9] But the brain does not have monopoly over what could be called organic consciousness, whose "content" is constituted by the organism itself or by its living elements. The term "content" should be taken here in the particular sense of "information content." What "informs" psychological consciousness (if we tentatively ignore kinesthesia) are the objects of the external world, their pattern,[10] which is transmitted more or less faithfully by sensory organs. What "informs" primary, organic consciousness, in contrast, is the form of the organism, its formative instincts, and the instincts directed toward a specific *Umwelt*. The brain does not bring the external world into existence as a world for the organism. But it allows the organism to act with detailed information on this *Umwelt* inherent to every living being.

Just as the brain cannot have monopoly over memory or invention, because it has itself to be invented or memorated by the embryonic being under formation, so it cannot have monopoly over consciousness. Consciousness—in the sense of "conscious perception of external objects"—has to be based on the brain's immediate consciousness of itself, of the brain as part of the living organism. Obviously the brain does not have internal sensory organs at its disposal to perceive, see, or hear what the acoustic or optical nerve brings it. We lack a third eye to see our occipital visual area and a third ear to hear our temporal aural zone. Ultimately, consciousness has to be united in an immediate way with the brain as living tissue for sensory consciousness to appear to be a property of the brain, an organ that is macroscopically arranged for sensory reception. Because the third eye or third ear is just a myth, like the "gaseous vertebrate," it is not the structure of the sensory organ or the structure of the brain as a macroscopic organ (like the stomach or the heart) that brings consciousness into existence for the first time. This structure only determines the way in which consciousness will be "informed" by the pattern of external objects.

Another order of considerations can help us mark the distinction between the brain as a macroscopic system for the use of organic consciousness and the brain as living tissue from which primary consciousness is inseparable. It is the case of automatic machines that contemporary industry is improving so quickly.

The rib cage, the stomach, the kidney, and the heart, as macroscopic structures, can be very easily imitated by machines, which may at times be used in medicine and surgery: an iron lung or an artificial kidney where the blood of an intoxicated person is circulated, and so on. No doubt these machines merely reproduce the massive operation of the organs in question. Nothing prevents us from perfecting them to the point of making them reproduce a "finer" operation: a true external artificial lung could theoretically replace not only the paralyzed rib cage but also the diseased lung by oxygenating the blood. Chemical indicators, sensitive to blood pH levels, would regulate the respiratory intensity of the machine, just like the reflex that normally excites the nervous center of breathing. These improvements would nonetheless come up against a limit as soon as we reach organic properties, like the faculty of reproduction, regeneration, and repair of functional wear and tear.

Similarly, the brain in its "massive" operation can be efficiently replaced by machines. This is not a utopia or a daydream; the "molar" cerebral functions of adaptation to the external world are in fact already aided and even usurped by various automatic mechanisms.

a. "Perception" takes place through a host of systems: floats, thermal circuit breakers, devices for magnetic "perception," ultrasonic probes, gyroscopes, radars, photoelectric cells. Notice that these artificial "sensory organs" can perceive or discriminate not only objects but also forms or, more exactly, patterns. A screen made up of photoelectric cells can distinguish not only spatial but also spatiotemporal forms, by taking into account the speed of variation of the phenomenon to be detected.[11] One day, we will certainly build garages equipped in this way, which will automatically open when their doors "recognize" the owners' cars. We have equipped the blind with radars. Theoretically, we can also imitate the mechanism of semicircular canals to balance automata. We can imitate the chemical mechanisms of smell and taste.

b. This perception is realized through servomotors and "servo" mechanisms (e.g., on ships, the subordination of a projector to the watchman's telescope, of the helm to the helmsman's wheel) and through relay switches.

c. A "guiding image" appears more and more frequently in automatic machines, a rough foreshadowing of a cortex capable of symbolic activity. A thermal screen—maintained by bypasses suitable for the same temperature as the engine, and tied on one hand to thermoelectric sensors and on the other to contactors—regulates its temperature. The operation of the Trappes railway is carried out by a ball-bearing mechanism: each ball, corresponding to a wagon, falls on a series of scales manipulated by electromagnets tied to stretches of the rails; the wagon, by advancing, controls the descent of the ball, which in its turn controls the railroad switch. The Vauban sluice gates in Strasbourg are controlled by hydroelectric boxes, where pressures on both sides of the lock are reproduced on a metallic membrane. Some companies of electric distributions in America use a *network analyzer*,[12] a scale model of their connections, which allows them to study the different problems posed by various accidents or unforeseen requests in the sector.[13] Electronic tubes (thyratron) tied to a mechanical "probe" of the scale model guide the work of the tool-machines, which can copy the model exactly, and so forth.

d. Anticipation and memory (in their mechanical aspect) are also not beyond the reach of industrial automatisms: timers; progressive orders at a predetermined pace; punched tapes; electrostatic or continuous-wave "memories" of calculators, which can be combined with the procedures mentioned earlier.

e. "Cybernetics."[14] *Giant electronic calculators built in America have recently attracted much attention. Enthusiast researchers like Wiener believe, not without reason, that the knowledge gained during the construction of these machines would help us understand cerebral operation in living organisms. There is no doubt that,* once installed *by the improvising consciousness according to certain connections and for a determined task, the brain can operate like an electric-relay machine, opening and closing neural circuits. A calculator's electric switches, operating on the binary number system, are similar to nervous synapses, which allow the influx to pass or prevent it, according to a law of all or nothing.*

Today we can build mechanical models of the brain in which a receptor part and a motor part can be distinguished, just as in the brain; in which the equivalent of a center of symbolic activity even takes shape; where breakdowns will be produced in one or the other of its parts, corresponding to agonsia and apraxia.

Evidently, this mechanical model of the brain as an organ of sensory or computational consciousness will not be a mechanical model of the brain as a living and directly conscious organ.

It is as legitimate and interesting to use the knowledge we have gained from the mechanical or electrical "brains" to understand the assembly realized by the conscious effort in the organic brain as it is absurd to believe that consciousness is only this assembly itself, in the passive and not active sense of the word *assembly.* The mechanical or electrical connections of automata, like the physiological connections of routine tasks, are just a projection of activity onto the plane of spatiotemporal operation. They are a set of substituted chains, substituted for improvised links of creative thought.[15] We rediscover within them the general schema of mental work: general task, perception of the particular problem, norms, recording and control. But these elements of the mechanical brains are nothing more than shadows. In the automatic brain, "control" is merely a control in the second or third degree. The norms are materialized by the assembly and do not dominate it.

For example, the way ENIAC performs division is typical: it subtracts the divisor from the numerator until the result becomes 0 or negative. Then it shifts to the next column (on the right) and adds the divisor until the result becomes 0 or negative. Thus, to divide 84 by 3, after having twice subtracted 3 from 8, it blindly continues to subtract 3 from 2 without being capable of survey and projection.[16]

$$
\begin{array}{r l}
3)84 & \\
-3 & (3\bar{2}) \\
\hline
+54 & \\
-3 & \text{ou } \underline{\underline{28}} \\
\hline
+24 & \\
-3 & \\
\hline
-6 & \\
+3 & \\
\hline
-3 & \\
+3 & \\
\hline
0 &
\end{array}
$$

Figure 20.

"Cybernetics" awakens today the same enthusiasms and the same illusions that were once provoked by hydraulic or pneumatic automata, from which Descartes drew his idea of the reflex. The schema of the reflex gave a remarkable impetus to nervous physiology and is still valid to a certain extent today, although it now seems completely subordinated. Calculators and reflex assembly obviously belong to the same order. It is no more extraordinary to use the knowledge we gained from the building of electronic calculators to understand cerebral operation than it is to use chemistry to grasp the mechanism of digestion or certain actions of hormones. The organic brain is already an assembly; it is carried out by the living being at each ontogenesis and according to a specific structure. It is perfectly normal for humans to remake, in the external circuit, systems that are auxiliary to the brain and, to some extent, analogous to it, just as they remake in external circuit a host of other organs or create tools that are auxiliary to organs.

We can design an automaton that sheds well-imitated tears when we tell it, "Your plea for clemency has been rejected." But then it will remain impassive when we say, "Take heart, the execution is this morning." If the engineer were to foresee the second sentence and *n* sentences that have the same meaning, there would always be an $n + 1$ that was not foreseen in the mechanism. We rediscover here the famous "telegram argument" invented by L. Busse and renewed by H. Driesch and

MacDougall. A man receives a telegram: "Your father is dead." His emotional and active reactions are considerable. If the telegram had read, "Our father is dead," with a single word altered, the reactions could have been entirely different; on the other hand, if the meaning of the first telegram had been transmitted in another language or verbally, the reactions would have been identical.

We also seem to rediscover Descartes's more celebrated expositions on "a machine that bears a resemblance to our bodies, which can pronounce appropriate words if we touch it in one place, but that cannot arrange words differently so as to answer to the sense of all that is in its presence."[17] It is nevertheless perfectly normal for the axiological "cogito" to lead us down a path parallel to that of the Cartesian cogito.

This encounter with Descartes and modern animists (Driesch and MacDougall) can help us clarify the thesis to which the facts point. Today it is clear that Descartes and the Cartesians of the seventeenth century poorly marked the break between what they called the "soul" and the body, or between what is better called "the domain of sense" and the domain of mechanical causality. There is sense and active finality in organic life. There are also mechanical systems in organic life. These systems are probably established by an active finality. But this finality subsists now in the fossil state; it is replaced by "chains" that function according to a step-by-step causality and that can be replaced by machines proper, created by human beings. The break is thus located inside the domain of organic life. It separates, in the organ, what is a massive arrangement and what is the living tissue capable of regulation. It separates, in the brain, what can be imitated by automatically regulated systems and what is thematic regulation and active finality. The soul, to use this term tentatively, or "primary organic consciousness," should therefore be deemed to act in every place where physicophysiological chains do not suffice to explain the total behavior of organs.

Let us consider the example of the heart. Suppose that a surgeon in the year 2000 managed to replace a failing heart by a self-regulating pump, with electric circuits playing the role that the sympathetic nerve and the vagus nerve play in the organism. If the circuits are disrupted, the pump will malfunction, whereas the in vitro cultures of myocardium fragments, in the absence of every vascular and nervous connection, present rhythmic contractions whose primary source resides in muscular centers surviving from the embryonic cardiac primordium: sinoatrial node, node of Tawara, bundle of His. Of course, physiologists think they can explain these rhythmic contractions in their turn by periodic chemical modifications of the myocardial cells, and it is indeed likely that chemical relays intervene. But in the end, it is necessary to arrive at a kind of mnemic melody, immediately inherent to the living tissue, combining its actions with the actions of secondary regulators and presiding over the play of chemical relays.

The case of the brain is identical, except that primary organic consciousness plays a much more considerable role here. It does not simply underlie the play of auxiliary mechanisms, intervening only in case of failure; instead, it dominates the operation of innumerable secondary systems, receptors, and effectors, directing them and improvising new links between needs and ongoing activities.

Another difference between "consciousness of the heart" and "consciousness of the brain" is that it is impossible and absurd to explain the totality of cerebral behaviors by physicochemical causes, because the "I" participates in them, whereas we can (implausibly but without absurdity) try to explain the action of the heart's embryonic centers in this way. An organ differs from a tool precisely because the border between the domain of sense or active finality and the domain of causality divides the domain of reality of the organ, whereas it leaves the tool in the domain of step-by-step causality, because the tool is built and inspected from the outside.

6

The Brain and the Embryo

Because they failed to posit at the start a reasonable hypothesis about the possible role of the brain in invention, memory, learning, and finalist activity, physiologists and psychologists were greatly surprised by the result of Lashley's experiments on the effects of cortical lesions on learning and memory.[1]

Lashley conducted his experiments on rats. When cortical lesions are introduced before learning, the deficit in the speed of learning[2] and in the level of performance measures the effect of the lesion on learning itself. When they are introduced after learning, the deficit measures their effect on memory. Lashley uses two kinds of trials: the trials of the first kind are *unskilled*[3] and do not require manipulations but are nonetheless quite difficult, because to reach the goal, the rat has to run over two pedals that control the opening of a door; the trials of the second kind require various manipulations: the door has to be opened by the rat itself, who has to lower a latch, pull a handle or a chain, or tear a strip of paper. Lashley wanted to know at a very basic level whether the rat's success would depend on specific cortical areas. To his surprise, the experiments showed that very considerable lesions—more than 60 percent of the entire surface of the cortex—were necessary to slow down and not to render impossible the learning of the two-pedal box, and more than 30 percent to slow down the learning of the manipulation boxes. His experiments showed, on the other hand, that the site of the lesion has no importance whatsoever. The slowdown in learning, which is nonexistent in small-size lesions, is quantitatively proportional to the size of large lesions, regardless of their localization. Still, the deficit seems to be due to a reduction in the animal's general health, exploratory activity, or sensory activity rather than to its intellectual capacities proper.

As to the memory of learning, experimentation shows that non-frontal lesions have no effect whatsoever. Lesions of more than a third of the frontal region of the cortex apparently abolish the memory of learning, but postoperative rats are capable of relearning the solution at a normal speed. It is not absolutely certain that a genuine memory loss is at issue, because similar experiments on apes (chimpanzees and inferior apes), in whom we can discern more easily than in rats the lesions that affect various areas (motor, premotor, or prefrontal) of the frontal region, show that paralysis rather than amnesia is at stake.[4] When paralysis in the wake of lesions of the motor area disappeared, through a restoration that is itself quite difficult to explain, the animal proved that it had retained the memory of the correct solution. Likewise, the lesions of the premotor area cause apraxias rather than amnesias: "The monkey can, for example, move toward the rope and grip it, thus showing that he knows it to be the solution to the problem. But sometimes he simply stands there and does not pull the rope. . . . In each case, what is lost is not what has to be done but the means that have to be implemented."[5] As to lesions of the prefrontal zones, they abolish not the memory of the solution but the right temporal serialization of acts that the correct solution requires. Experiments on apes are all the more interesting because their cortex took (relative to other parts of the brain) almost as great an importance as the cortex has in humans and because we cannot invoke the intervention of subcortical mechanisms in learning and memory as we can do for rats. Moreover, the experiments on lesions of the subcortical regions in rats generated results similar to those obtained on the cortex: in this case as well, the disorders are proportional to the quantity of damaged brain tissue.

These results are not isolated oddities. In all domains of cerebral and nervous physiology, the observations and experiments that have been accumulated for years, especially in Germany and America (Bethe, K. Goldstein, von Monakow, Jordan, Carmichael, Child, Coghill, and P. Weiss), reveal very similar phenomena. The cortical surface—and the same holds, more or less distinctly, for all nervous centers—does not operate as a material surface with geometricophysical properties. It in no way operates like the screen made up of photoelectric cells we imagined earlier and in which one part would obviously not be equivalent to the whole. Through the cortical surface, signifying themes are transformed into schemas of action (anterior frontal and motor cortex), or inversely, sensory patterns evoke significations (posterior cortex).

In principle, themes and significations cannot be localized. They are, as we have observed in our summary description of signifying action, outside the plane of space-time, at least the plane of space-time understood as a set of juxtaposed elements that exert step-by-step actions. Physiologists as well as psychologists had to recognize the thematic character of learning relative to the terminal nervous effectors. The animal we condition spontaneously generalizes the stimulus, and it has to be progressively taught *not* to generalize. Far from being reserved for humans, this faculty of generalizing and transferring is universal among living beings. "On this point," C. T. Morgan notes, "we can say that no essential change took place in the course of phylogenesis." All the experimental works of animals psychology[6] have observed it: "In the most primitive beings as well as in the highest in organization, psychology starts off with the operation (the generalization) that we consider to be the most complicated."[7]

Instinct is always thematic. Instinctive behavior is not stereotyped; it is made up of chains of behaviors, and the sensory sign that triggers it is never akin to a distinctly outlined key, because the experimenter can always deceive the animal with approximate forms. Very often, it is even the need for an absent form or object that sets the animal's activity in motion. This activity ceases when it has approximately created or located this object or form: a nest, a den, a spider web, a sexual partner, and so on. The conformity of instinctive action to the description of senseful and finalist activity is obvious. The animal reacts to an absence just like the traveler who rides the train because he is not where he wants to be. To speak of gnosia inscribed in the constitution of the nervous system is to propose an a priori implausible hypothesis, because it is difficult to inscribe an *absence* of stimulus in the nervous tissue. Furthermore, studies concerning the effect of cortical lesions on the maternal and sexual behavior of the rat have been undertaken (Beach), and the effects were completely analogous to those of Lashley's extirpation experiments. To produce deficits, extensive lesions (more than 20 percent of the cortex) are necessary. They are proportional to the quantitative significance of the lesions; and it is less the instinct that appears to be affected than its means of induction and execution, just as, in Lashley's experiments, it is less the memory of learning that is affected than the means to be implemented.

For a long time, the reflex was treated as though it were the element of an assembly similar to that of a photoelectric screen. To be sure, there remains some measure of truth in this conception; with the reflex, we are in the domain of nervous effectors, and the spatiotemporal "step by step" begins to reign once again. It is only more significant that experimentation also revealed thematic rather than spatial unities in this case: "The simplest spinal reflex 'thinks,' so to say, in movements, not in muscles," in terms of functional utility.[8] By relying on natural experiments (brain diseases and battle wounds affecting the brain), Goldstein insisted, on the other hand, on the significant fact that the reflex always stands against the backdrop of general adjustment of the organism as a whole, that it is like a "figure" in a visual field.[9] We can no doubt interpret these traits in terms of synaptic connections: Gasser, for instance, used this strategy to explain the reciprocal inhibition, which is indeed a particular case of the general phenomenon described by Goldstein. But we still have to understand how, according to the momentary needs of the organism, it is this action that becomes a "figure," by thematically controlling the synaptic openings and closures through suitable chronaxic change or another procedure. It is very striking that nearly all reflexes can be designated by a psychological rather than physiological name—"stretching," "scratching," "recovery," "support," and so forth—and that physiologists are forced to classify reflexes according to their end rather than according to the nervous means used.

Last, experiments of electric excitation of the cortex, especially of the ascending frontal (motor area) zone and of the premotor zone situated further ahead, have unequivocally shown that, when they are possible, the relatively precise localizations are localizations of *themes* of movement or action and not localizations for the control of this or that muscle. The exact same thing takes place when we electrically excite sensory areas, assuming the trepanned nonanesthetized patient can describe his impressions.[10] As one would expect, on account of the point-by-point projection of the retina, the electric excitation of the striatal area (the visual area proper) affords sensations of brilliant light; these sensations are localized in the higher part of the visual field when the lower part of the striatal area is excited, and vice versa. But when the electric stimulus is applied to the neighboring area (Brodmann area 18), the patient experiences meaningful visual impressions; he sees flames, stars, shiny balls, butterflies, various objects, and even people. The lesions of the same area do not produce cortical blindness like the lesions of the striatal area but rather visual agnosias, just as the lesions of the premotor area do not produce cortical paralysis but apraxias. We should add that even very extensive lesions of the visual area cause no substantial deficit in the memory of discrimination between visual forms. Provided a small part (1/10) of the area remains, the patient is like a man who, instead of disposing of a whole mirror, has no more than a tiny fragment of a shattered mirror: he is very ill-at-ease, but with compensatory movements, he can continue to see all that he could see with the whole mirror. His brute impressions are no longer the same, but the senses are conserved, just as the sense of figures that can be observed in a mirror fragment—provided one learns to look for this sense—is the same as the one that is reached through the whole mirror.

We had to frame Lashley's experiments in this way before trying to interpret them. The totality of known facts allows us to conclude that the impossibility of strictly localizing the functions of the brain or the nervous system is always tied to the thematic or finalist character of action and perception. It is a priori implausible to interpret what Lashley calls cerebral equipotentiality or the equipotentiality of the extended cortical zones—namely, the startling fact that a part of the brain or of a sensory or motor zone is equivalent to the whole—through a mechanical model in which a step-by-step causality reigns.

This equipotentiality runs parallel to embryonic equipotentiality, which the facts of twinning, regulation, and regeneration allowed us to postulate for a long time and which many laboratory experiments since Driesch's works have clarified. A fertilized egg—and even, in countless species, the blastula and the young gastrula—is not a mosaic of territories that are irrevocably destined to engender this or that organ. For convenience's sake, embryologists distinguish "presumed primordia" in the young egg or embryo (for instance, in a young *Triton gastrula,* the animal hemisphere contains the primordium of the epidermis or the neural primordium), but the presumption simply indicates the normal outcome of these territories.

In a typical experiment, Spemann (1918) sections the largest part of the animal hemisphere, turns it 180 degrees, and places it on the vegetative hemisphere, interchanging in this way the neural primordium and the epidermal primordium. After cicatrization, the embryo continues its development without abnormalities: the epidermal primordium generates the nervous system and the neural primordium generates the epidermis. It is by reflecting on experiments of this kind that Lashley could say, in jest and with some exaggeration, that one sometimes has the impression that if one could remove the whole cortex of the rat and place it back on the brain after having turned it 180 degrees, nothing would change in the animal's behavior.

Carried out later, the operation would fail. The prospective value of the two territories (neural and epidermal) would no longer be the same. At a certain moment, the two territories undergo a "determination" that remains invisible for some time but that is soon translated by an apparent differentiation. Presumption, determination, differentiation: the three stages have to be clearly distinguished. The presumption concerns only the knowledge of the biologist, who knows what takes place in normal development; but Spemann's experiment proves that the animal hemisphere in the young *T. gastrula* is equipotential before determination. After determination, the equipotentiality is conserved, but only for the more limited, determined territory. New experiments of cutting and rotation of a piece of the territory can now only succeed at the interior of this territory. Sometimes the determination is only active for certain axes and not for others. If we turn the bud of a limb 180 degrees, if we shift it from the right part to the left part of the organism, or vice versa, it can develop regularly according to its new position. But sometimes it also retains its own anteroposterior direction, which is determined before its character as left foot or right foot or before the dorsal–ventral direction.

Thus embryonic equipotentiality, like cerebral equipotentiality, is bound up with the thematic character of development. The cascade of determinations has a thematic character, because determination precedes differentiation and differentiation proceeds in its turn through themes that can only be designated with abstract terms: a limb bud is determined as a "foot" (as "foot" in general) before it is determined as right foot or left foot. It is also tied to its finalist character (in the strictly etymological sense of the word) because, in all the extraordinary regulations allowed by equipotentiality, the normal end is reached despite the operative disruption of conditions, materials, and means.

The proof that the parallel between Lashley's experiments (on the regulations of behavior in the wake of brain lesions) and those of Spemann (on the regulation of organization in the wake of embryonic lesions) is not artificial is this: the progress and maturation of behavior in embryos and young animals follow laws similar to those of organic development; they go from the global reaction of large muscular, not-yet-innervated groups to individualized and differentiated reactions of muscles with specialized innervation (Graham Brown, Coghill). In Coghill's observations and experiments on *Amblystoma,* the locomotive behavior has first of all a global manifestation: the organism takes the form of a C, then an S, with the flexion propagating from head to tail.[11] Only then do the feet develop and progressively participate in the movement. The theme of the body's sigmoidal movement thus precedes the differentiation of the limbs' locomotive reflexes.

We can therefore say that, in contrast to the irreversibly differentiated organs of the adult, the brain and more particularly the cortex retain some measure of the equipotentiality of the egg or of the embryonic territories. For the brain, anatomic differentiation is not accompanied by a physiological differentiation in the broadest sense of the term. The evident homology between the two equipotentialities completely rules out the idea that cerebral equipotentiality is a kind of secondary effect, obtained through the complex network of neural interconnections and so-called association fibers: the embryonic territory is equipotential, and yet it does not possess such a network. It would be implausible to explain equipotentiality, which is so similar in both cases, by a cause that would only be present in one of the two. Instead, it is logical to admit the conservation of a primary property of the living organism for both the operation of the brain and the primary rhythm of the heart. The association fibers probably have nothing to do with cerebral equipotentiality. They have sufficient use as instruments of transport for sensory modulations and motor effections. Quite the contrary, their presence and their anatomic differentiation contribute to restraining equipotentiality with "rail" effects [*effets de rail*].

Starting from this perspective, we can understand a surprising fact noted by all specialists of the nervous system. Even in cases where it seems that the organism would benefit from obtaining a point-by-point transmission of patterns (e.g., in the transmission of retinal images to the striatal area), it can be said that the organism destroys its own work, or renders it systemically more difficult, by complicating the network of direct fibers of projection with innumerable synaptic fibers of interconnection. These fibers can only render the transmission diffuse, and it would have been easy to keep it anatomically precise by not forming neurons or association fibers. As we know, the cortical areas have in principle a simple architecture of the same type. They comprise six layers, whose relative importance varies depending on the area and of which some appear more particularly devoted to radial conduction (the layer of giant pyramidal cells of the motor area is the best known) and others (more particularly those on the surface of the cortex) to lateral association.[12]

The schema of these areas as well as of the retina, which has a very similar structure to a cortical area, can be figured as follows:

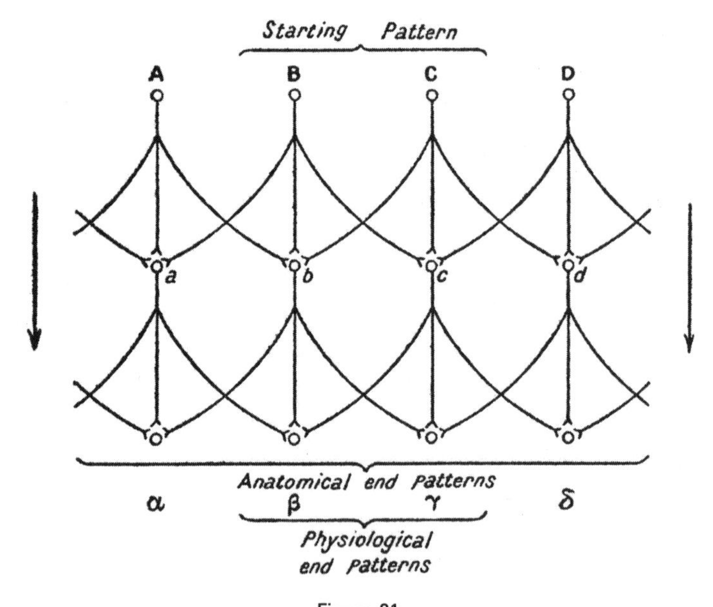

Figure 21.

It seems that this structure is meticuously designed to render the strict projections of ABCD on to $\alpha\beta\gamma\delta$ impossible. In fact, the organism corrects this curious anatomy with complex and poorly understood physiological procedures: summation (e.g., if B and C alone are excited, the neurons *bc* and $\beta\gamma$ receive more influx than the neurons *ad* and $\alpha\delta$, and the resistance of synapses can be vanquished for $\beta\gamma$ and not for $\alpha\delta$); temporal summation (Lorente de No); chronaxic accord or discord (Lapicque); reciprocal inhibition (Gasser), and so on. Whether the organism goes about it in one way or another, the precision of the nervous operation is generally very superior to what we would expect based on the anatomic structure. But if it is true that cerebral equipotentiality is independent of association fibers, it is by contrast perfectly normal for the brain, the instrument of a thematism that cannot be reduced to a chain of step-by-step causality, to operate according to physiological rather than anatomic relations. In reality, physiological connections can be established or broken with little energy. By contrast, anatomic connections would transform the brain into a pure machine or an irreversibly differentiated organ like the liver or the lung, in which embryonic equipotentiality is from then on "expended" in immutable structures. Psychic assemblies for a determined task, which the intelligent act uses but to which it is not reduced, would correspond to definite structures; they would be unable to constitute the auxiliaries of thematic behavior. Behavior would then become nothing more than the operation of anatomic connections.

Association fibers are thus neither the necessary nor the sufficient condition of equipotentiality. The physiological wiring of neural components corrects the preset nature of these neural networks. Far from constituting the condition of equipotentiality, a preset neural anatomy would in fact hinder it.

In principle, equipotentiality is absolutely indifferent to the existence or nonexistence of association fibers. These fibers affect only the method of reception or effection. The foveal part of the retina differs from the rest of the retina in that there are *fewer* lateral synaptic interconnections between the elements of radial conduction: each cone seems to be in contact with a single bipolar cell. If the interconnections were the key to equipotentiality, then foveal vision would have to be stripped of all "thematism," something that obviously contradicts the facts. In the case of a visual learning, the foveal vision is more precise by its technique than general vision, although it is no less capable of "equivalence" and "transfer."

The existence of association fibers—provided, of course, their closure or opening is controlled physiologically and is not anatomically irreversible—does not hinder equipotentiality, any more than it conditions it. The association fibers that unite the occipital lobe to the temporal lobe probably allow some conjugated behaviors, but they do not effectively mix and mingle visual sensations and auditory sensations as ingredients. They remain as distinct from one another in consciousness as the optic details are distinct from the foveal vision.

We observe the most striking contrast between the enormous intellectual deficits of dementia, where it is often impossible to detect the least macroscopic brain lesion, and the negligible or even null effects (on intelligence proper) of enormous lesions, not only of sensory areas, but of the frontal lobe, the presumed seat of symbolism-based behavior. The effect of the now-common operation of prefrontal lobotomy is to bilaterally section a part of the fibers that unite the prefrontal lobe to the thalamus. Comparative trials done before and after the operation do not find any substantial intellectual deficit. Probably due to an improvement in the affective state (decrease of tension and of affective anticipation, not of explosive affectivity), some trials were even more successful (Freeman and Watts).[13] Even in cases of apparent deficit, we realize on closer examination that the reduction bears on the aptitude of symbolicoaffective "assembly," on emotional perseveration, rather than on intelligence itself (Spearman's *g* factor). After the operation, the lobotomized patient remains just as capable of "eduction of relations" or "eduction of correlates," which, according to Spearman's thesis, represents general intelligence and, we might add, the typical thematism of cerebral equipotentiality.

The *ablation* of frontal lobes naturally engenders serious deficits but does not cause dementia or radical amnesia.[14] Unilateral ablation is tolerated without marked psychological disturbances; bilateral ablation destroys (much more severely than lobotomy) the capacity for symbolic assembly and for behavior that is seriated according to a plan. It destroys the auxiliaries that are indispensable to the action of the *g* factor, but as far as we can tell from a small number of cases, it does not destroy the *g* factor itself.

The experiments of bilateral lobotomy on apes (Bianchi, Jacobsen) lead to the same conclusion. The "assembly," the *set*[15] indispensible to deferred reactions, is disturbed: the animal is very easily distracted from a task it has to retain in its consciousness. More easily perturbed is the architecture of a plurality of "assemblies": the animal is incapable of planning its behavior. It can no longer organize its behavior in a well-defined series of actions. Because it is very difficult to experimentally distinguish between animal intelligence proper and an auxiliary as essential to intelligent behavior as the retention and seriation of tasks, it is difficult to reach a conclusion on the role of prefrontal lobes in intelligence and memory.[16] But, at any rate, it seems that the facts do not support the assertion that the frontal lobes are the seat of intelligence. It is enough to browse through the accounts of Köhler, Guillaume, and other experimenters to realize that nearly all the problems of intelligence posed to animals implicate a correct seriation of tasks. Some privileged cases allow us to distinguish, in the reasons for the animal's failure, between what is due to a lack of intelligence and what is due to an emotional accident that demolishes the edifice of "assemblies" for the undertaken task (e.g., the failure of Chica as a result of a noise that frightens it and makes it reverse its efforts in an absurd way).[17] Theoretically, then, the problem of the exact role of frontal lobes in intelligent behavior is accessible to experimentation, even if the limited number of experiments has not been entirely conclusive. It may now be assumed that the frontal cortex is in the service of consciousness, memory, and intelligence but that it is not, in its general structure and in the architecture of its association fibers, a kind of instrument for becoming intelligent. That dementia is produced much more easily by disturbances that affect neural cells (intoxications, various degenerations) than by massive lesions of the brain indicates at least that cerebral equipotentiality, as it appears in intellectual thematism, is due to the nature of the living tissue rather than to the massive architecture of the brain as an organ designed for perception and action on the external world.

Just as cerebral equipotentiality should not be explained by neural connections, which are on the contrary in its service, so embryonic equipotentiality cannot be explained by physicochemical factors. We can be brief on this subject because we have discussed the question at length in a previous work;[18] when discussing genetics later on, we will come back to the logical impossibility concealed in theories like that of Th. Morgan and de Dalcq, which claim to explain the embryological formation through the action of a substance or a chemical on genes. So-called chemical organizers, spilled out by genes or by embryonic organizing centers, are ordinary chemicals. They can be construed as triggers or, rather, as invokers of formative psychomnemic themes, which are summoned by them to pass into the plane of space-time. But it is simply absurd to turn them into the *causes* of the complex structure of the invoked organs. These substances act in the manner of smells that invoke memories in humans or instincts in animals; they put the embryo in circuit with mnemic themes that, once "invoked" (determination), pass into the actual (differentiation). The chemical theory of organization is doubly absurd, for even if we concede that an ordinary chemical can be the cause of a complex structure, it cannot be its cause in the manner that is in fact typical of embryonic development. This development always goes from the abstract to the concrete.[19] Determination and the initial differentiations can only have abstract expressions: axis of symmetry, dorso-ventral axis in general, cephalic region, caudal region, somites in general, limb buds. As psychological observations demonstrate, this is how the sexual instinct invoked by chemical means (hormones) goes from the abstract to the concrete, from an undifferentiated stage to a more differentiated one, in a development similar to the one that orients the progressive change of tubers and primitive genital folds in male or female organs or that transforms into a hand or a foot the primitive palette where the future fingers are, in the beginning, buds similar to one another.

So, strictly speaking, embryonic equipotentiality is not a "property" of material tissues and their chemistry, any more than cerebral equipotentiality is a property of the material cortex and its interconnections. The general notion of equipotentiality designates the fact that the area at issue can be put in circuit with this or that theme, relative to which it is still indifferent,[20] and that the theme can gain a foothold indifferently on this or that large or small part of the (embryonic or cortical) area. Relative to the tissue, equipotentiality represents something negative, not positive; and the notion was very muddled by the expressions "real potentiality" and "total potentiality" that were often employed simultaneously with it. Real potentiality designates what the territory really becomes and total potentiality all that experimentation (on other individuals of the species) shows that it can become. The equipotential territory is not *at once* itself and something else; it is *not yet* what it will become when it is put in circuit with this or that mnemic theme.

At a given moment, I have a memory that "occupies" my consciousness (and my brain activity as well); but I could easily have had another if a different "invoker" had intervened to put me in circuit with it. This does not mean, however, that my consciousness and my brain contained both and all the others "potentially." The Aristotelian expression is meaningless in this case and, contrary to what Driesch believes, has nothing to do with Aristotelian notions. What is at stake is a possibility of placement-in-circuit with various themes that are not in space-time.

This symmetrical argumentation about the primary character of equipotentiality in both cases—against the connectionist hypothesis and against the physicochemical hypothesis—is reinforced by its very symmetry. If the connections or gradients of substance were not simple accessory means, it would be difficult to understand how embryological and cerebral phenomena could exhibit such kinship.

These means do not play the same role. The "connectionist" means in neural organs are subordinated to the tasks of reception of sensory patterns and motor effection, which lead to well-localized terminal paths. The embryo's physicochemical means are subordinated above all to the tasks of coordination, synchronization, and distribution in development. In the egg or the embryo, which is at first totally equipotential—except in species where a very precocious determination masks this character—the determination distributes this equipotentiality into more limited territories, which develop from then on with relative autonomy (as revealed by the experiments of delayed graft, where the transplant develops "unintelligently" according to its origin, *herkunftgemäss,* and no longer according to its new place). In normal development, the chemical means intervene to guarantee, spatially and temporally, the coordination of territories.

Precisely because it is a matter of an auxiliary technique, accidents can take place, which lead to results similar to those of delayed grafts in experimentation. These accidents are called monstrosities, and E. Wolff has compellingly shown that experimental embryology illuminates the majority of natural monstrosities. Just as the ordinary psychological memory as well as the instinctive memory can be scrambled with abnormal signals, so normal development is scrambled by accidents in physical or chemical relations between the territories. For instance, if we damage a localized region of the embryo, which, as a result, undergoes an arrest in development, "this elimination entails as a secondary outcome the coming together of territories that are not contiguous in normal evolution."[21] If the damaged region is axial, the two lateral primordia will enter into contact, and if at the moment of contact they are determined (thematically) without yet being differentiated, they will fuse together in the same way that two adjoined urchin eggs can fuse into a single egg. This is the case of Cyclopes and Symeliens (monsters with only a single lower limb). Equipotentiality, which should have been "distributed" into two equal territories, remains undivided. The case of experimental or spontaneous cyclopia and symmelia weighs heavily in favor of the primary character of equipotentiality, because it is for lack of an axial organ interposed as an obstacle that they spontaneously play out and lead to a single organ. Spontaneous or experimental half-double monsters, shaped as a Y or a lambda, illustrate a perfectly analogous phenomenon and prove that the mechanical or chemical means of numerical individualization come only after a unity that is given first. It would be contradictory to explain this unity by the very means that, on the contrary, limit or distribute it. The monstrosities prove in their own way the finalist surveying unity, whose effort is deceived by causes that, it is tempting to say, "are independent of its will." The monstrous embryo is never an arbitrary form: "The malformations are secondary modifications of a primordium that is first constituted according to the normal mode."[22]

In the order of behavior, the "means" (the neural connections) cannot provoke the same kind of accident as physicochemical means in ontogenesis. But like all techniques, they can provoke other analogies, as when a ring in the chain of means is falsely called upon and triggers meaningless results. To find examples ad libitum, it is enough to open, not a treatise on psychiatry, since psychogenic or humoral disturbances are certainly much more frequent in this domain than neural disturbances, but a treatise on neurology: paresthetic sensations, synopses, and synesthesia; illusions of amputees; hyperalgesia; various agonsias and apraxias; auras; hallucinations; deliriums; nonpsychogenic and nonhumoral anxiety or euphoria, and so on.

Despite the clear contrast between the two great means in the service of embryonic equipotentiality and cerebral equipotentiality, the organism has to pass from one to the other in the course of development. A chemical triggers the formation of the neural plate, and the nervous system then uses the neural connections and switches [*aiguillages*] in its operation. As we know, the organism never abandons the first means, even in the domain of behavior, because in the adult, the hormonal regulation that prolongs the action of embryonic organizers is essential not only for physiological operation but for psychological life itself. The pituitary gland is a nervous organ at the same time as it is a gland with internal secretion. The sympathetic system functions in a semichemical way by releasing adrenaline, acting more or less diffusely; it is thus an invoker as much as a trigger. Although their general thematism has certainly nothing to do with substances or spatiotemporal structures in the organism, the majority of instincts are subjected at once to humoral conditions and to neural conditions, and we can deceive an instinct or render it "monstrous" by acting on either of these conditions. We can deceive an animal by appealing with a trick to instinctive gnosia that—without being materially inscribed in its nervous system and even without presupposing its integrity—presuppose all the same a certain operation of sensory nerves (artificial light to accelerate egg laying, decoys and snares for hunting and fishing, and mannequins for artificial insemination are examples), in the same way that we can deceive an instinct with humoral alterations and transform, in their behavior as well as in their organization, genetic roosters into chickens with estrogen or genetically female guinea pigs into male guinea pigs with testosterone.[23]

But researchers have managed to grasp the facts more closely and to uncover in some privileged cases the passage from chemical means to connectionist means. If we graft, for example (Harrison, Detwiler, Weiss), a batrachian limb bud in an abnormal position, the grafted limb seems to draw toward it the medullary nerves that normally innervate it. If the grafted bud is already "determined" with respect to its axes, its direction of bending after growth will be *herkunftgemäss*, the original direction, and if we graft it upside down, its bending will be "absurd" relative to its new place: one foot will bend when the other (indigenous) foot bends, but it will bend in reverse relative to the host organism. This proves that whether it is upside down or in the right position, the grafted foot—in becoming functional—attracted the nerves that correspond respectively to the extensor and flexor muscles; these latter were already determined as such embryologically by the induction they underwent before transplantation (very likely, a chemical induction). We are thus forced to conclude that *the neural connections follow in a docile way the quality of the muscles to be innervated.* This is what P. Weiss expresses picturesquely when he says that everything takes place as if each muscle knew its name (muscle name theory).[24] The neural connections appear at their rank in the cascade of determinations; they are produced according to the general theme and the proper sense of organs to be innervated. The chemical inductors or organizers serve to regulate the distribution of the themes of development; the neural connections obey in their turn the induced quality of the territories, to the point of persevering in the accidental or experimental monstrosity and rendering it definite.

In such cases, we uncover the passage from theme to "consolidated"[25] structure. If the adult organism seems at times akin to a machine and in fact functions partially like a machine, it is a machine that has constructed itself. Obviously this self-construction cannot be understood unless we start from a kind of self-survey of the machine's structure at all of its stages. Self-survey is, as we will see, another way of denoting and defining equipotentiality. The industrial construction of the steam engine rests on human consciousness, the domain of self-survey whose objective manifestation is the equipotentiality of the cortex, which extends embryonic equipotentiality. The mechanical "step-by-step" bonds of the machine rest finally on the primary self-bonding of the embryo. Chemical means and connectionist means in internal-circuit organization, industrial means in external-circuit manufacturing: these subordinate techniques of organization and behavior presuppose a primary mode of unity by self-survey. This mode has to be clarified; but we are now able to say that it cannot be understood through assemblages of causes acting step by step.

We will not spend much time on the "explanations" of equipotentiality proposed by biologists, embryologists, and neurologists. They are equally worthless. It is typical that at times the same author employs all of them successively, thus proving that he is satisfied with none. Lashley, for example, evokes them all.

a. *Quantitative explanations.* There is an ordinary "equipotentiality" in countless adult tissues: we can live with a single lung, a single kidney, and even with a fragment of a lung, which is in this sense equivalent to the whole. Because the result of Lashley's experiments is expressed quantitatively ("the performance deficit is proportional to the quantity of the damaged cortex"), the explanation seems to emerge quite naturally from the facts: the case of the cortex would be similar to the case of the pulmonary or renal tissue; the cortex would act "massively." But this is obviously pure wordplay. The equipotentiality of the pulmonary or renal tissue has nothing to do with the brain's equipotentiality. The effect (oxygenation or chemical purification) is directly measurable. In contrast, a behavior or the solution of a problem is not itself quantitative. The illusion stems from the fact that the deficit of a behavior can be estimated by indirect means (the time required, the number of errors, etc.).

The weakness of so-called holistic biological theories or of the numerous *Ganzheittheorie* proposed recently derives precisely from their conflation of the whole as "quantitative mass" and the whole as a domain of forms capable of self-survey (and thereby of equipotentiality). There is in reality nothing extraordinary about the qualitative equivalence of the part and the whole in a quantitative mass. We can sweeten a glass of water with one or two pieces of sugar; the "holists" sometimes seem to believe, as a result, that there is nothing strange in the fact that a single blastomere engenders an entire embryo, that a single cerebral hemisphere accomplishes the same work as the two hemispheres or a fragment of a sensory area the work of the intact area. It is clear that the quantitative theory sidesteps the following problem: an adult organism is structured as a whole, a sensation or a behavior is equally structured in this way; how can an overall structuration be independent of the spatial support in which it is realized? Half a piece of sugar is still "sugar," because in sugar nothing more than an indefinitely repeated molecular microstructure is at issue. But half a vehicle is not a "vehicle." The striking thing is that half a newt or urchin gastrula is sometimes not only still "newt" or "urchin" but a whole newt or urchin. Similarly, half an auditory or premotor zone is at times an integral instrument of complex and structured sensations and behaviors. With damaged retinas, I can have "vision"; and we are used to estimating the residual vision to one decimal point. But the problem is that I recognize the same forms when I see them with different or more or less extended parts of my retina. The contrast with what a photoelectric screen would produce is the heart of the matter.

b. *Global psychological or physiological explanations.* Very different in appearance, they are similar in fact to quantitative explanations. They appeal to a general factor such as "vigilance" (proposed by Head). Physiologists who object to the use of a psychological notion in this situation can always replace it with "scientific" equivalents, such as the threshold of excitability or any general factor. But just like the first explanations, these ones also fail to address the essential structural problem. It is perfectly possible that the notion of vigilance responds to something significant. Every human being has the experience of states of mental stupor when he sees without seeing; a cortical lesion can determine this state of general stupor fairly well. Yet this stupor, or this increase in the threshold of excitability, or this loss of mnemic availability, can only mask the effects of equipotentiality. It is not enough to appeal to its opposite (vigilance) to have a positive explanation of equipotentiality. Let us add that we do not see how this explanation would apply to embryonic equipotentiality.

c. *Pure Gestaltist explanations.* They have had enormous success in both embryology and psychology, precisely because they seem to tackle the right problem of the conservation of structure—no longer the conservation of a constitutive microstructure but of the structure as a whole—despite the quantitative decline in the material support for this structure. If I cut in half a magnet, a soap bubble, or a capacitor charged with electricity, I would still structurally have a magnet, a soap bubble, or a capacitor in which electric charges are distributed in the same way as in the original capacitor. This time, the case is apparently similar to the cutting of an egg, a gastrula, or a cerebral area in half. But as we have seen, the analogy is more ostensible than real. According to the principle of the least action, the Gestalt-form that conserves itself results from a dynamic equilibrium operating step by step and leading to very simple, homogeneous, and maximally symmetrical structures. As P. Weiss,[26] Humphrey,[27] and others have stressed, behavior often progresses toward symmetry, homogeneity, and harmonious and well-connected lines. But it would be truly difficult to see in the progress of organization from the egg or the blastula to the adult a progress toward symmetry and homogeneity. The regulations after lesions and the various regenerations take place through complex reshufflings and through the intervention of various tissues, which produce appropriate neoformations, or through buds of regeneration, along with the work of specialized cells that had first to migrate to the right place. We need a great deal of good faith to believe that they can be explained by a spontaneous dynamic rebalancing like that of electricity in a capacitor. In general, the transfer of a behavior or a habit after lesion is a much more complicated phenomenon than a simple transfer of forms onto a reduced material. It is almost always accompanied by appropriate and significant qualitative modifications. In an old experiment of Lashley (reported by Humphrey), a rat trained to cross a labyrinth that included left turns underwent an operation that made it impossible for it to turn left.[28] The rat nevertheless succeeded in crossing the labyrinth by turning three-quarters of a turn to the right, which replaced a quarter of a turn in the opposite direction. Lashley's postoperative rats are occasionally forced to cross the labyrinth by crawling on their front feet, by performing somersaults, and so on.

d. *Pure connectionist explanations.* They are practically ruled out by the very nature of the problem; and in any case, they cannot be applied to embryonic equipotentiality, unless the preformationist explanations in embryology are equated with the connectionist explanations of cerebral behavior. The facts cited previously against pure "Gestaltist" explanations are much more decisive against pure connectionist explanations. If a pure Gestalt transfer poorly explains the creativity of the rat who makes three-quarters of a right turn to replace the quarter of a left turn, neural connections established by learning would not explain it at all: "None of the studies of learning or retention of the mazes after brain lesions has given the slightest indication that the maze habit is composed of independent associational elements. There was never amnesia for one part of the path with retention of another."[29] When a mother sends her child to deliver a letter to the neighbor, she expects him to go through the backdoor if the front is closed. By the same token, a sheepdog gathers the flock at the signal of his master by taking into account the arrangement of the sheep and the nature of the terrain.[30] In these cases, Humphrey notes, beyond the various terminal connections implicated in the performed action, there has to be a more general neural pattern that activates the particular patterns of actions.

e. *Explanations by "Gestalt" and connections.* Virtually every author falls back on a combination of the two preceding theses. The general pattern Humphrey refers to would be a Gestalt-form, dynamic and transferrable, which could thus set varied connections into action depending on the circumstances. By combining two bad theses, the author hopes to generate a valid one. Equipotentiality would mean a two-stroke regulation: (1) a simple dynamic regulation, of the type of a physical regulation (through the establishment of a gradient or of a self-distribution[31]), would (2) entail a change in the effectors used. This dualist theory can be adapted for the explanation of embryonic equipotentiality: it is enough to replace the neural connections with the genes that are thought to contain the explanation of the structures. In this case, the gradients of the chemical substance provide the general pattern. Depending on the local level of concentration, the genes that are triggered at different thresholds engender this or that organ. When the experimenter cuts a *T. gastrula* in half along the sagittal plane, the gradient regulates itself at first like electricity in a capacitor. Then the affected genes generate, according to new thresholds, other organs than those they would have produced, with a similar overall form but different dimensions (Child, Dalcq). The great success of gradients in animal and especially plant biology stems from the fact that they explain the adaptability of organic formations.[32]

General pattern (Gestalt)	Chemical gradient
Determined neural connections	Genes

Figure 22.

Koffka and Lewin adopt this dualist theory under the name of the theory of the "circular process." When an animal approaches an attractive prey or flees a danger, the neural effectors that enter into play can be very varied, but they are controlled by a simple dynamic situation: by the increase or decline of tension resulting from the approach or moving away of the danger or the goal. The "circular processes" are analogous to feedback systems whose significance in modern machines as well as in physiology has been underscored by cybernetics. In these systems, an obtained first effect reacts on the subsequent effection, which thus takes the result into account.[33]

Finally, Lashley also adopts this dualist theory. Consider, for instance, a relative discrimination learning in rats. The animals learn to react positively to the most brilliant or the greater of two circles, regardless of their absolute brilliance or magnitude. Pure connectionist theories already fail to explain this learning. They also fail to explain the conservation of this habit after a lesion of the visual cortex. But if we suppose (Figure 23) an S gradient along which an equilibrium between the Ps and Ns, *whatever they may be*, is established, it will be the global equilibrium on the S-line that determines the activation of the R-path or the L-path and not the fact that the stimulus is P rather than P' or P".

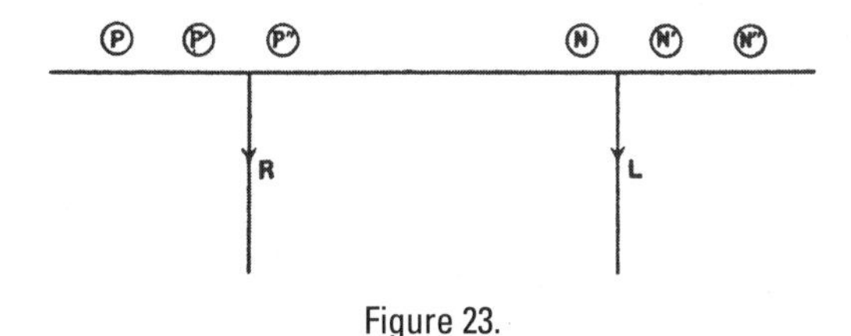

Figure 23.

These explanations fare no better than the preceding ones. We have already shown at which point the genes in embryology are incapable of fulfilling the enormous task assigned to them.[34] The theory of the circular process contradicts the facts. The animal that flees or approaches does not resemble a body that obeys a difference of potential in a field. Köhler's own observations (and those of Guillaume and Meterson) on chimpanzees clearly show that the intelligent solution struggles against the direct attraction of the goal, but not like a force that can be represented by a vector.[35] In a detour, the part of the trajectory that distances the animal from the goal has to reverse the "circular process." So, to explain why the animal persists in this provisional direction, we have to assume that the parts of the trajectory that distance it from the goal are balanced by the parts of the trajectory that bring it closer to this goal, in the same way that the ascending part in a siphon is balanced by the descending part. But unlike what happens in a siphon, in the animal's detour, these parts of the trajectory are crossed *at different moments*. Cerebral dynamism, if it exists, is thus altogether different from a physical Gestalt dynamism where the forces that balance one another are all equally present. Something in the field of consciousness or in the cortical field of the animal has to temporally and spatially survey the entire trajectory for the dynamically opposed parts of this trajectory to be maintained because of their sense *(meaning)*[36] relative to the ideally projected and not-yet realized total path. The *(finalist)* direction of the trajectory has to struggle against the direct dynamic tendency to move in the (vectorial) direction of the goal. This is even clearer in the case of a more complex behavior. When a chimpanzee nibbles an overly wide board to thin it down and turn it into a stick, in what way can a "circular process" explain this long and arduous action? How can a "circular process" explain a political or military human maneuver that is intended as a bluff or a provisional concession? Here the notion of dynamic equilibrium is just a metaphor of doubtful interest. An equilibrium between senses *(meanings)* differs radically from an equilibrium between physical forces. This latter equilibrium has a unique outcome in which the constitutive vectors are combined. In contrast, an equilibrium between senses *(meanings)* keeps distinct the various senseful actions that have to be composed, and this is why these actions can be ordered in a series.

This critique applies to Lashley's schema. The schema is vaguely plausible only for the particularly simple case of discrimination between relative degrees of brightness. And one must not examine it too closely, because a discrimination (like a trajectory with detours) implies that the animal keeps distinct the terms to be discriminated and does not operate like a scale whose needle only indicates a *difference* in weights. Besides, it must be that Lashley was not satisfied with this explanation, because he ultimately replaced it—while remaining within the framework of dualist theories, but abandoning the identification of the general pattern with a gradient or a dynamic equilibrium[37]—with a curious "wave theory" of learning: the cortical surface is equated with a continuous network (association fibers being so numerous that the influxes can practically diffuse in all directions) like the surface of a pond on which stimulating or inhibitive wave trains propagate, trains that conserve some measure of the stimuli's form while moving in the cortex and combining with other wave trains.[38] The figures formed in this way can be instantaneous or stationary, like those of the vibrating plates sprinkled with sand and excited by a bow. This wave model would show that the cortical responses do not depend on points stimulated in the receptive areas and, on the other hand, that a fragment of the cortex can function in the same way as the whole.

Because this theory would not in any case account for embryonic equipotentiality, we need not discuss it. It is doubtful that it constitutes an improvement over its predecessors. If the pattern of stimuli is conserved intact in the wave trains, then is there a significant difference between this theory and the connectionist explanation? If the cortical waves modify this pattern, they can only render it rougher but cannot extract its thematic character, and the difference between this theory and the hypothesis of dynamic rebalancing would this time be insignificant. Lashley seems to forget that the essence of equipotentiality does not reside in the *circulation* of forms from one point to the other in the nervous system but in the *thematic equivalence* of forms. This theory's main interest lies in showing the hopeless state of the deterministic theories of equipotentiality.

7

Signification of Equipotentiality

There is, nonetheless, some measure of truth in the dualist theories, as we indicated in chapter 5. The duality of the nervous system is real: as receptor and effector system on one side, with its anatomic dispositions and its physiological availabilities, and on the other, as living equipotential tissue and, as such, in relation, like every living tissue, with the domain of spatiotemporal senses. Because positivist scientists refuse to recognize this domain, they try to fabricate its equivalent on the plane of anatomy and physiology, naturally without success. "Positivist" scientists do not seem to realize that this failure is very fortunate for the unity and intelligibility of the facts and, thus, for the intelligibility of science. Let us suppose—contrary to all possibility—that one of the preceding theories, or a new related theory, proves to be absolutely true. Everything would then become objectively clear in the neural operation, even the most paradoxical facts revealed by Lashley's experiments. Everything, except the role of consciousness. We would no longer understand what consciousness does in the real world, precisely because we could completely explain the neural operation. We would find ourselves in the same situation that we were in at the time of Huxley and Maudsley. Back then, it was thought that the neural operation could be explained by ready-made connections and, therefore, that consciousness had no longer any use, like the cogs that, in the famous story, the watchmaker realizes he forgot on the table after having reassembled the watch, which works perfectly.

That it is impossible to explain cerebral equipotentiality "causally" should seem entirely natural. This equipotentiality marks the site where a coupling can take place between the cerebral system, as a system, and the world of consciousness and thematic senses that make use of this system, not like a pianist uses his keyboard but in a much more subtle way that we have yet to study. A man's consciousness, memory, and ideas do not constitute a spiritual "second man" (avatar of the "gaseous

Vertebrate") who is superimposed on the first man of flesh; they constitute a proper domain that can be regarded, in a first approximation, as distinct from the "observable" cerebral system. Both because of their analogy and because behavior is indissociable from organization, we are forced to treat cerebral equipotentiality and embryonic equipotentiality in the same way. The observable embryo, like the observable brain, cannot represent its entire reality without contradiction. Like the brain, it is in a relation with a domain of memory and signifying themes, which take hold of it and dominate the visible structural transformation. Numerous signs indicate that these two "inobservable" domains are one. The organic memory that guides the differentiations of the embryo, the organic inventions that perfect the species in the course of successive ontogeneses, closely resemble psychological and individual memory, consciousness, and the faculty of invention. The same effects are owed to them; their borders are very fluid: what is a tool in certain cases (the product of psychological consciousness) is an organ in other cases (the product of organic consciousness). The brain is an embryo that has not finished its growth; the embryo is a brain that begins to organize itself before organizing the external world.

The first peculiarity of the brain is to be in relation with the domain of themes and senses, not only directly, like the embryo, but indirectly, by means of the external objects it perceives and creates. In the adult organism, the brain is an area that has remained embryonic. It remains connected to the inobservable domain of senses, while the rest of the organism, having finished its growth, retains some contact with this domain only to the extent that it cannot be reduced to pure "substituted mechanisms," to the extent that mnemic themes and rhythms, carried over from the embryonic state, continue to "inspect" these innumerable machines. The second peculiarity of the brain is that its differentiations are reversible, whereas the differentiations of the rest of the organism (except in certain inferior organisms) are generally irreversible. Placing the adult brain in circuit with mnemic themes or original senses entails only a provisional closure of the cortical network's synaptic connections, which are always physiologically open. At the moment of a definite perception or action, this closure transforms the brain instantaneously into a "finished" organ, that is, a differentiated organ like the others. A living being can only perform a single task at a time in the order of behavior. It can only have a single "assembly" for a given task. If, contrary to all possibilities, it could spend its entire existence in the same assembly, the cerebral connections that are thus closed for good would be comparable, in their immutable anatomic structure, to the connections of the renal or pulmonary tissue, which always carries out the same chemical work. In reality, the living being ceaselessly passes from one action to another. Ever-new thematic systems, controlled by mental or psychic laws and not by a physiological causality, alter at every instant the "closures" of the neural network, and this amounts to transforming the network into an organ with a new structure. A given psychological recollection or idea only mobilizes the brain provisionally. The brain is quickly available for another differentiation.

In contrast, the mnemic themes that are successively summoned in the course of embryonic development determine an irrevocable differentiation. Primitive embryonic equipotentiality thus disappears progressively; it is distributed in more and more limited areas. The theme of organs, by taking shape, ceases to be a theme to become a structure. The finalist sense of the organ remains obvious, but this sense is incarnated or fossilized, in the same way that the theme of invention in a machine built by an engineer is replaced by substituted mechanical links. Relative to the embryo that he was, the adult realizes in a sense the ancient myth of the divinity that is transformed into a laurel tree. To be an organ for the creation of organs is what equipotentiality allows. This definition makes clear the resemblance and the difference between the fertilized egg or the young embryo and the brain. Both respond to this definition. To say that the young embryo is like a brain at the moment when a recollection starts to emerge is not a metaphor. It is impossible to interpret the facts highlighted by experimental embryology (anteriority of determination over differentiation; *ortsgemäss* or *herkunftgemäss* development of grafts; induction with regulation, etc.) through mechanical or dynamic models.[1] Only the "psychological model" of mnemic initiation can account for the facts. And it is not absurd to match a domain of primary consciousness to the observable embryo, just as we spontaneously match a consciousness to the observable head or brain of a living being. A senseful theme, which the observable structure expresses but does not exhaust, can only have a subjective existence.

There is no reason to imagine that this primary consciousness or subjectivity of the "determined" young embryo is vague, confused, or psychoid rather than psychic, à la Hartmann, Becher, Bleuler, or psycho-Lamarckians. The precise or vague character of a consciousness can only be inferred from the structure of the systems or behaviors it establishes. Embryonic systems and behaviors are miracles of subtlety and precision. No doubt, the theme "foot" or the theme "lung" or "kidney" is abstract at the outset, as we have stressed; but an abstract idea is not a vague idea. P. Weiss called his theory the *muscle name theory* probably in jest and thought he was offering a pure metaphor, but he touched on the crux of the matter. The metaphor only bears on the word *name*. Obviously, the embryonic muscle does not know its "name" (as extensor or flexor), but it certainly knows its own nature; it knows its own sense, if not its signification. The embryo's primary consciousness is no more vague than the consciousness of the adult; it has another direction, it "concerns" itself uniquely with the organs it is constructing. A worker absorbed in his labor forgets the rest of the world. This certainly makes the external world vague for him, but not the object on which he labors—quite the contrary. As the organic labor is pursued, primary consciousness, at first equipotential, seems to lose itself in the more or less automatic structures it establishes. The worker seems to disappear in the work. The distribution of equipotentiality, which allows the division of organic labor, is no doubt accompanied by a distribution of primary consciousness, since a graft transplanted after determination develops "mindlessly" according to its origin and not according to its new place. But even if this distribution has to correspond to a dissipation of primary consciousness, it cannot be understood as a passage to the vague state. In a sense, the "I" of adult consciousness is a specimen of this "distributed" primary consciousness, because it is not tied to the entire organism but to the nervous system and particularly to the cortex as an organ of behavior. "I-consciousness" is not "vague" relative to its own tasks. It is "in the dark" relative to organs other than the brain and relative to the brain itself, as irrigated breathing organ or as seat of various chemical phenomena. But nothing allows us to suppose that primary consciousness, distributed to other organs, is more confused relative to these organs than the "I-consciousness" relative to sensations and behavior. For these primary consciousnesses (e.g., for the mnemic rhythms of autonomous cardiac centers), it is our "I-consciousness" that would appear as confused and "psychoid-rather-than-psychic."[2] Our "I-ideas" are very vague relative to the operation of our organism. Let us be fair, then, and not expect too much of the ideas of our primary organic consciousness vis-à-vis external behavior, which is the affair of the "I." The discovery of blood circulation was a feat on Harvey's part; the admiration for Harvey gives the measure of what the admiration for the "I-consciousness" in general must be. If our heart and our arteries had the time to judge our brain, they would not have a very high opinion of its capacities: it has taken so many centuries to become aware of what happens a few decimeters from it and even within it.

What may mislead us in this question is that the "I-consciousness," tied to the brain, only receives the communications of organic consciousness in the form of instinctive drives, often peremptory but always imprecise and "protopathic." The "I-consciousness" is thus inclined to attribute to total organic consciousness (when it believes in it) the same traits it attributes to its communications. A sexual drive, for example, is no doubt vague for the "I-consciousness" that is not aware of its nature: "I don't know what's happening to me," exclaims Cherubino.[3] Yet it is quite natural for the communication between distributed consciousnesses to have a confused character, which none of the distributed consciousnesses shares. The sexual drive in the "I-consciousness" is confused, but the formative instincts—to speak like von Monakow and Mourgue—that had to make the male or female gametes according to a rigorous "machining," that had to construct not only the sexual organs but also the very complicated physiological cycles that allow hormonal and neural sensitization, are necessarily quite precise. It is simply absurd to believe à la Schopenhauer and von Hartmann, followed by contemporary psycho-Lamarckians, that these formative instincts are a sort of degraded consciousness or an unconscious will. In all likelihood, the sexual sensitization of the nervous system and of the "I-consciousness" results from procedures like the one through which, during embryonic development, an already determined area induces in its turn the determination of a neighboring area by means of a chemical. It is the endocrine condition of the blood, rather than the neural incitations coming from organs, that determines the eroticization of the nervous system. Each area is equipotential in itself, and to this equipotentiality corresponds a precise consciousness. But the influential passages from one area to neighboring areas are translated by a confused impression in the influenced area, until the invoked mnemic themes differentiate consciousness at the same time as the organs or behaviors at issue.

Cherubino quickly becomes wiser. From the vague and atmospheric impressions induced by the madeleine, Proust reconstructs the immense edifice of his recollections. Touched by the optical vesicle, the ectodermic tissue rapidly constructs a lens and a cornea from a simple thickening of the cephalic epiblast.

These three facts are perfectly equivalent. In each case, it is a matter of the passage from one domain of equipotential survey to another:

1. either from one embryonic area to another
2. or from the organic domain to the psychological domain (instinct)
3. or from an mnemic sphere (closed onto itself) to the "I-consciousness"

In the evolution of species, the nervous system is at the primitive stage very rudimentary. And it is rightly concluded that the psyche in the ordinary sense of the term (i.e., consciousness turned toward adaptation to the external world) must be equally rudimentary. Whether one is or is not a strict behaviorist, little matters here: animal psychology finds that the development of behavior and of the psyche follows quite faithfully the development of the nervous system. But the common error consists in extrapolating rashly and in believing that where a nervous system is lacking, a consciousness must also be lacking. The second-consciousness (i.e., turned toward the external world) of an annelid or an echinoderm must be undeniably more vague than that of a rat, a monkey, or a man. But if the absence of every nervous system corresponds to the absence of every secondary consciousness, it does not necessarily correspond to the absence of primary consciousness, which is tied directly to the organic form and not to the form of the nervous system. Nor does it further imply the vague character of this primary consciousness. If, by going from man or chimpanzee to annelid, we go from a precise secondary consciousness to a vague consciousness, nothing allows us to assert that, by moving from echinoderm to protozoan or to plant, we move to an even more vague consciousness, ε if not zero. On the contrary, the facts imply that we continue to find a consciousness of another kind, primary and perfectly clear and precise in its kind, albeit turned toward biological organization and not toward the world.

An infant who has just been born has only a confused consciousness of the external world. This consciousness quickly takes shape as he ages. By retroactive extrapolation, we have a tendency to believe that before the moment of birth, this consciousness was nonexistent or evanescent. And yet the embryo is capable of thematic behavior. It is therefore more logical to assume that the embryo had a consciousness of a different nature, with a different content, but as subtle and complex as its behavior can imply, which is saying a good deal.

We do not know how to go about persuading our reader, if he had the patience to follow us, that we do not inject this thesis with any fantasy or any metaphorical vagueness.

Like all half-truths, panpsychism has done more harm than good. It is panpsychism, more than behaviorism, that prevents us from defining primary organic consciousness clearly and precisely, because it "fills the place" with a secondary consciousness in the infinitesimal or diluted state. The harm can be traced back to Leibniz and his "tiny perceptions." Understood in this sense, panpsychism is as false in the psychobiological order as would be in physiology a thesis that, having vaguely glimpsed the fact that assimilation and respiration are cellular and not merely macroorganic phenomena, concluded that there has to be in each cell tiny stomachs and tiny lungs. Whereupon biologists, failing to locate these tiny stomachs and lungs, would be tempted to deny every cellular assimilation and respiration.

The facts of equipotentiality should set us once again on the right track. Equipotentiality is the objective functional aspect that a particular mode of reality assumes for an observer: a consciousness—that is, as we shall soon see, an absolute form, an absolute self-surveying domain. As the assembled and interconnected structures of a machine are the sign of a consciousness that was once applied to this assemblage and represent, so to speak, fossil finality, so equipotentiality is the sign of an actual consciousness. The adult equipped with a brain was at first an embryo without a neural plate. The embryo's primary consciousness is therefore primary from all points of view relative to the consciousness that is turned toward the world. The "I-consciousness" as a domain derives from the domain of embryonic consciousness. If we want to grasp the facts, we have to become used to dissociating consciousness and brain and to associating consciousness and organic form. The brain is not an instrument for becoming conscious, intelligent, inventive, or reminiscent. Consciousness, intelligence, invention, memory, and active finality are tied to the organic form in general. The brain's "superiority" or its distinctive character is that it is an incomplete organ, an always-open network, which thus retains equipotentiality, the active embryonic consciousness, and applies it to the organization of the world.

It remains for us to study more closely the content of the intimate relation between organic form and consciousness. For if we have noted how everything leads back to this close relation, which certainly contains one of the most pivotal secrets of finalist action, we have not yet tackled the problem in itself.

8

The Reciprocal Illusion of Incarnation and "Material" Existence

The examination of the facts forces us to rethink the Cartesian break between a thinking soul and a mechanical body. The opposition, as it emerges from recent observations and experiments, is instead between (1) the organism as a set of tools or a set of organs insofar as they are tools and (2) consciousness (primary or secondary, organic or cerebral) that assembles multiple elements in such as way as to turn them into "amboceptors" in a causal chain and that thematically oversees the operation of organic machines, regulates them in case of lesion or failure, and thus gives to organic structures the property of equipotentiality.

This dualism, as different as it may be from the Cartesian variety, is still a dualism and seems to pose the same problem that tormented Descartes's successors: how can two types of existents as heterogeneous as consciousness and the body be intimately joined together in the unity of the living being? Up to now, we took it upon ourselves to pass ceaselessly from the perspective of consciousness and the subject to the perspective of the body and the object. We have to justify these passages by rediscovering unity, or a certain unity.

It often happens in the history of the sciences that a problem that at one point appeared insoluble seems to solve itself. Such is the case here. The solution was discovered several decades ago by many authors,[1] and we attempted to formulate it precisely in an earlier text (*La conscience et le corps,* 1937). Heymans, whose metaphysical works we will set aside, had expounded it with perfect clarity in various articles (gathered in *Gesammelte Kleinere Schriften,* vol. 1, La Haye, 1927), but he committed a gross error that we will come back to. On the other hand, in the particular domain of psychiatry, Adolphe Meyer had protested for a long time against the abrupt distinction between body and mind, somatic theories and psychological theories in mental medicine, and called for a psychosomatic medicine.

This solution can be formulated in a few words: the problem posed by the duality of consciousness and the body, consciousness-organism and body-organism, is illusory for the excellent reason that there is no body. The "body" is the byproduct of the perception of a being by another being. The perceived being is perceived by definition as an object, in the etymological sense of the term. It appears as independent of the observer, and this leads him to substantialize it. This substantialized object is then called a "body." But we have to consider several cases.

a. A and B are two humans observing each other (Figure 24). A's reality for A, or B's for B, is the totality of his cerebral and organic consciousness, with organic consciousness distributed more or less in cellular or noncellular subindividualities. A's reality for B appears in B's cerebral consciousness as a perceived object, which B will call A's body, and vice versa. Because humans are social beings, A quickly adopts on himself, for normal use, the viewpoint of the observation of objects and not of pure *self-enjoyment.*[2]

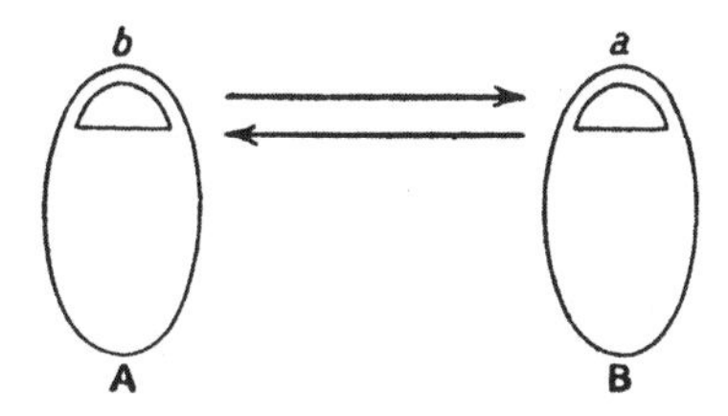

Figure 24.

Of course, it must not be entirely forgotten that A is above all a center of conscious activity, though young children and dogmatic behaviorists do forget this. Moreover, A adopts the dualist point of view on himself and, by analogy, on B, on all other humans, and on superior animals: he is consciousness and he is body. This illusion is even more natural because, independently of every social relation, man is constituted in such a way that he can be in a relation of observation or even in a "social relation" with himself. He sees his arms and hands extended before him, and he can speak to them like Lady Macbeth; he sees most of his body when he is seated or when he examines himself in a mirror. But the fact remains that if it were possible to conceive a human being living alone, without a mirror, with an immobilized head, incapable of looking at or touching himself, we do not see how such a being could have the curious idea of considering himself as double and as composed of consciousness and of a material body. If he were endowed with philosophical reflection, he would quickly notice a certain duality between his active "I-consciousness" and the more passive states of consciousness (suffering, malaise, euphoria). He would suspect that a hierarchy and a distribution exist in his conscious being, but this duality would not at all resemble in his eyes the old consciousness-body duality.

b. A observes a tree and no longer another human B. The same illusion takes place. The tree is perceived as an object. This time, the analogy no longer forces A to attribute to the tree (as he attributes to himself) a self-enjoyment doubling its objective appearance. He hastens to consider the tree as a pure body without "interior doubling," without a subjectivity of its own. If A is a biologist, he will analyze the operation of the tree's organs, but without any of the reservations that even the most obdurate materialist must experience when he examines a child or an animal. And yet it is clearly unacceptable to consider the tree as a pure body without a subjectivity of its own. The tree-object exists only in the perception of an observer A, and the tree as a pure body is merely a substantialization of this tree-object. The real tree grows and develops as a unity; it maintains its own form. It does not depend on the accidental perception of humans or animals passing in its vicinity. Nor does it depend on the biologist's observations. The careful examination of the facts can lead us to suppose that the tree's unity is not as clear-cut as the unity of an animal. A young oak or a young horse chestnut has, for instance, leaves as large as those of an adult tree of its species; from this fact we conclude that the tree is a colony of organs rather than an organism proper. But the more or less unitary modes of subjectivity have nothing to do with the general necessity of presupposing an "autosubjectivity," a "for-itself," in the plant. The plant is subjectivity and not body, just like the animal.

c. A is a biologist who, by means of a now-possible technology, observes the occipital cortex of B, who in his turn observes the tree. A sees nothing in B's cortex that resembles a sensation or an image of a tree. But if he electrically excites a certain region of B's occipital cortex, B will have a distorted and modified vision of the tree—at least, this is the most likely outcome, according to a direct inference from analogous cases. And indeed, this reaction can be a proof that the real cortex in itself, at least at a certain level of its bonds, *is* the subjective and conscious sensory field, and that it is this self-enjoying[3] field that appears to the observer A as a gray or white substance or as a "physiological state" of this substance. If A observes in succession B's cortex and the tree that B is observing, he will see in both cases nothing but bodies. Because, unlike B, the tree does not speak and cannot describe its impressions, and because the observer has to be quite attentive and intelligent to interpret the relevant and finalist regulations of the plant as indirect signs of subjectivity, A will be tempted to see in the tree only a pure body that is subject exclusively to the laws of classical physics.

If A were to observe the cortex of a dissected corpse, the observed appearance would not differ substantially from the appearance of a living cortex. And yet experience shows that this time the observable structure deteriorates quickly. This is the proof that a dead cortex does not enjoy the same kind of internal bonds as a living and conscious cortex and an excellent confirmation of what the observation of the living brain suggests: some of these bonds are the very consciousness of the observed man.

d. A observes a cloud. Because this cloud has no self-subsistence of its own and takes very varied forms at the whim of meteorological conditions, it is no longer necessary to suppose that it has a proper subjectivity as a cloud. By contrast, the question arises for the water molecules that constitute it, because these molecules have a subsistence and a form of their own. It even arises for the step-by-step bonds between these molecules, which constitute the instantaneous unity of the cloud as a physical phenomenon. If A were to see a wave moving over a pond, he would be tempted to consider the wave as a body. A more attentive observation would reveal that the droplets of water rise on the spot. The wave is therefore just a phenomenon, and the question of its proper subjectivity no longer arises.

e. A observes a machine. He examines its structure and its operation. This machine has a unity, yet clearly not a proper unity, because it results from the play of amboceptors assembled by the engineer and because, for want of maintenance and inspection, the machine quickly reverts to a state of scrap metal. No doubt, in both this and the cloud's case, the metal molecules or atoms have to be deemed (until further consideration) to have a "for-itself" of their own, because they actively retain their form and their unity in the absence of any external maintenance.

f. A observes a man who is less fortunate than Aurelia's fiancée;[4] he lost his arms and legs and even some internal organs but was reconstructed by a very advanced surgery, thanks to Plexiglas "props" and to automatic machines that replaced his organs. Obviously the man's artificial part should be viewed in the same way as the steam engine. In the normal organism, the parts of organs that are made up of dead cells—like the nails, the hair, the enamel of teeth, and so forth—have a self-subsistence only through their physicochemical components and their step-by-step bonds. More generally, their "macroscopic" organic operation is just a play of amboceptors.

We have examined a sufficient number of cases to be able to draw general conclusions. There is no body, that is, no material body, whose existential status is exhausted by the fact that it is purely and simply a body, massive and extended, without any subjectivity of its own. Mass and extension, spatiotemporality, dynamic and geometric properties of bodies cannot be true "properties"; they cannot belong inherently to beings observed as bodies but only to "autosubjective" forms or forces, if we can use this strange term. *Matter* and *material body*: these terms do not designate a kind of particular *stuff*,[5] supposedly different from a *mind stuff* or a domain of consciousness. Every real possesses itself; otherwise, who would possess it?

As B. Russell notes, the distinction between mental and physical (in the sense of "material") "belongs to theory of knowledge, not to metaphysics."[6] Russell is right in this sense: it is the "mode of apprehension" of the real B by the real A that makes the real B appear as a body or a material object. But we should also speak of *observation* and not of *knowledge*. I can know B's consciousness (by sympathy, empathy, and analogy and, above all, by the unity of beings in the unity of a sense) without transforming this consciousness into a body. But I can only observe it as a body. And it is easy to account for this. Observation is a physical event, whereas knowledge is a mental act. A observes B, or the tree, or the cloud: this boils down to saying that his retina is the seat of the impact of photons that emanate from various elements of B's structure. If, instead of vision, we were to turn to another sense, the observation would always be ultimately reduced to an energetic interaction. For observations proper, a photographic plate or a similar laboratory instrument can replace the sensory organ, often with advantage. If it were true that experimental science can be essentially reduced to a series of "index readings," as Eddington claims, by eliminating all the "inobservables" (in the sense that Heisenberg and Jeans give to the term) as much as possible or by leaving them to the domain of inferences, then it could be said that science observes and does not know. In truth, science does not relegate to common sense and to realist metaphysics the care of transposing the observed into an intuitive image of the world. It is equally realist, and it approaches observations with images of the real or "comprehensive" mathematical schemas. The discipline of "possible observation" is no less necessary to scientific knowledge; it gives this knowledge its distinctive character.

In everyday life, sensation is at once, indissociably, observation and knowledge, a physical event and an act of knowledge. It is a physical event insofar as the sensory organ is a system that can in principle be replaced by an artificial device; it is an act of knowledge insofar as the living tissue of the organ or of the corresponding cerebral area—or rather what appears as organic tissue to an external observer—forms part of the equipotential and autosubjective domain that is the very reality of the knowing being. Sensation is an act of knowledge and not of pure observation, insofar as it is the act of a being already in the world, capable of grasping significations and of having a sense of the "other"—a sense as primitive as the intuition of its own existence. Pure observation would never be knowledge, but only event, exchange of energy. Pure knowledge would remain virtual, because it would provide no details about the "other." It is the combination of observation and knowledge in sensation—in other words, of the living being's primary, autosubjective, organic consciousness and of physical events on the sensory organ—that allows a "detailed knowledge" of other beings. In radio emissions, the carrier wave is a physical reality as much as the modulations added to it. In sensation, only the "modulation" is physical, and the "carrier wave" is the primary subjectivity supplied by the living organism. Because the modulation alone provides the content of information about the external world and all the details of knowledge, one spontaneously overlooks the rest, all that is autosubjectivity, as much in the observer as in the observed. Moreover, common sense—without reaching the materialist or behaviorist purism of science, which tends to transform all objects, including humans, into bodies or into pure physical phenomena—is materialist for all the beings incapable of manifesting their interior life. Humans who lack imagination are "Malebranchists" about inferior animals and plants. "It doesn't feel," Malebranche said about his dog. We are all "Malebranchists" about physical reactions.

These considerations provide the key to the distinctions that should be drawn between the various bodies or between bodies and step-by-step phenomena. Everyday language employs the same term, *body,* to designate the observable organism of a man or an animal and to designate a structure or a mineral cluster. The same is true in French and German (the words *corps* and *Körper*). Language is justified, as we have seen, by the general nature of observation: just like a machine or a cloud, a living organism is observed only as a structure that emits photons. Whatever this structure's own mode of bonding may be, nothing of it appears in the pattern of light waves it emits and in the pattern of the photoelectric effects produced on the perceptible surface. The bonds are always inferred, never observed. Nothing is easier than leading these inferences astray. A wax figure at the museum, an automaton, the shadow of a person on the cinema screen, easily creates an illusion. Whether I perceive the circular appearance of a planetary nebula, a rainbow, a solid metallic sphere, a soap bubble, or an amoeba at rest, I always see a circle; and yet the modes of bonding in these various cases are extremely different. Once they participate in my "perceptive space," and provided they have the same structural appearance, the most heterogeneous figures become mental images characterized by the conscious domain's mode of unity. They are thus subjected twice to a treatment that confuses them: first, all of their proper bonds are suppressed in the pattern of light waves; then, they all participate in the mode of unity of the consciousness that perceives them. Experience, induction, and the extended observation of forms, of their operation and of their behavior, have to intervene to distinguish these forms. Even spontaneous experience easily distinguishes between aspect-forms and other forms and quickly refuses to consider the rainbow and the wave over a pond as "bodies." Enlightened common sense considers them, not as bodies, but as phenomena that owe their unity to the continuous and statistical action of a law. But the distinction between the other types of bodies and between their mode of bonding is much more difficult, as the history of science shows. Superficial observation—and even thorough observation—of the movement of celestial bodies does not enable us to determine whether they are held in place by guardrails or by the solidity of the crystal spheres, whether they obey purely dynamic relations (attraction at a distance), whether they follow a geodesic of non-Euclidean space-time, or whether they are divine spirits that obey the principle of the best. The superficial observation of the human body allows us to distinguish it from a wax figure and to distinguish a living being from a corpse. But even scientific observation does not allow us to easily distinguish the human body and its behavior from an automaton and its operation (in fact, the nondistinction is affirmed by Watson's disciples) or to distinguish it from a dynamic Gestalt-form of the "soap bubble" type (this indistinction has also been affirmed, and it is even a recent discovery). Even in psychology, paradoxically enough, researchers have not reached a definite conclusion about the mode of bonding that constitutes mental unity; and they believe it is possible to borrow from the type of bonding of external bodies the model that would allow them to elucidate all psychological phenomena: ancient atomists explain knowledge by shocks among atoms; associationists speak of the attraction between images considered as things; "Gestaltists" apply not only to the body but also to the mind explanations that rely on dynamic, step-by-step bonding according to an *extremum* principle.

When we observe a being sufficiently large and complex that its structure can be reproduced on a perceptible surface (a tree, the mobile letters of an LED board, a living or a dead cortex), we are exposed to all the errors of what Whitehead called a "misplaced concreteness" and to all kinds of uncertainties about this being's modes of bonding. But extended observation (which reveals the behavior of the being and not simply its instantaneous structure), experimentation, and induction allow us in principle to perform the necessary discriminations. A hierarchical structure, a unified behavior, self-regulation and especially self-repair, equipotentiality, the observable criteria of teleology (as a behaviorist like Tolman can define them) lead us to imagine modes of bonding that differ completely from the step by step, which can explain the subsistence or the operation of an aggregate, a Gestalt-form, or a machine. The self-regulation of a Gestalt-form can be explained by side-by-side interactions according to extremal laws. The self-regulation of an automatic machine can be explained by the arrangement of its pieces and its amboceptors, which propel one another. In any case, because we observe no form in its "subjectivity," no bond in itself, there is no reason to believe that we are scientifically obliged to reduce everything to step-by-step bonds. Nor is there any reason—under the pretext that nothing is a body—to assign a subjectivity to what is merely a cluster, an aggregate, or a mechanical assemblage.

Heymans commits this mistake when he claims to go to the end of what he calls his psychic monism and renews Fechner's dreams about the soul of the earth, considered as a psychological individual.[7] *This mistake derives from a more serious and more fundamental error. Heymans does not distinguish between the observed beings' various modes of bonding. He admits that physical laws are merely the reflection of a concealed real causality but models this real psychic causality on the physical, step-by-step causality. He simply replaces physical determinism with a psychic determinism of the same type and is therefore incapable of distinguishing between a pure aggregate, like the planet Earth, and an equipotential system, like the brain or the embryo.*

Let us imagine that A observes from above, no longer a man B, but a large crowd of human beings marching in an immense procession on a congested street or gathered in the central square of a city. If A observes from a considerable distance, he may not know that he is observing human beings. He notes that this crowd, this material "fluid," behaves "mindlessly" and unpredictably. If the head of the procession comes up against an obstacle, the queue continues to advance and to squeeze against the head, producing a kind of pressure surge. If the head starts to walk again, a sort of wave-decomposition gradually propagates toward the queue. If the crowd exits the central square, it flows frictionally through the available tissues at a calculable speed. In short, the laws of fluid mechanics account for the observed movements much better than the laws of individual psychology. At times, orders emitted by a loudspeaker modify the play of these altogether physical laws by acting directly on the conscious individuals. A becomes aware of this error, like the physicists who discover the primary laws beneath the statistical laws. Yet, by and large, the shape of the crowd is well determined by the fact that the movement of each individual is tied to the movement of others only "step by step." And to this extent it is not only useless but surely false to speak of a soul or a consciousness of the crowd, as if it existed as a distinct being capable of autoconduction and finality. A crowd of very intelligent men bears a striking resemblance to a crowd of stupid men or even animals or molecules.

This example shows that the negation of the body or of matter as a distinct entity does not entail the affirmation that an autosubjectivity resides behind every "object" or phenomenon. The molecules that make up a cloud, a machine, or the Earth can have a subjectivity as much as the humans who make up a crowd. But the crowd, the machine, the cloud, and the Earth do not have one. "Physical existence" designates a mode of bonding between elements, not a category of beings. If the interactions among the components are superficial in nature and propagate step by step, we will be fully entitled to speak of physical existence, even if each of the components is mental or intelligent. It is difficult to define at this point our concept of superficial interaction, because we have yet to define interaction in general. But we can tentatively think of a shock between corpuscles that retain their individuality or an action of pure power or pure pressure between humans who treat one another as simple obstacles or simple means, without bothering to persuade one another, or who treat one another as a "human material."

This distinction between physical "body" (obeying a step-by-step causality) and organic "body" (unified and capable of equipotentiality and autoconduction) avoids of the dialectical distinction that Hegel renewed between the "in-itself" and the "for-itself," the in-itself supposedly primitive relative to the for-itself. We cannot help but detect here a reflection of the old metaphysics, which are the products of "prescientific" conceptions of the world. The in-itself is the *Grund* of the old German philosophy and even of the primitive Chaos of theogonies. In France, the notion has in particular evoked memories of the old materialist mechanism. On a primitive, blind, and deaf Ground, consciousness, the "for-itself," establishes itself and alone gives a sense to this "primitive." "It creates the world by naming it."[8]

The facts do not confirm this poetic interpretation or these scientific dialectics. Psychological consciousness in the ordinary sense of the word, which is specialized in the sensation of external beings through the special arrangements of the cortex and of the sensory organs, is not the only real "form." Every being, every center of activity, is its own subject and possesses itself. Every being that is not an aggregate, every "organic" being in the broad sense in which Whitehead uses this term—which also includes the individualities of physics and chemistry—is a form, that is, directly self-possession, "for-itself" as well as "in-itself." Brute, blind, and deaf existence has to be understood *starting from* this presence of forms that possess themselves, in the same way that the laws of classical physics can be rediscovered from the data of microphysics. They derive from these data by virtue of the multiplicity of beings which, having become foreign to one another, only touch by their edges, superficially, and only act on one another step by step; they can thus form clusters, processions, or crowds incapable of autoconduction.

How does this multiplication of beings occur? We do not claim to know. Yet biological facts, and even chemical facts, attest to this double operation without disclosing its secret. On one hand is a multiplication that remains dominated by a surveying unity and that retains equipotentiality (the cellular multiplication leads to the development of a multicellular being from a unique cell); on the other is a multiplication that leads to a multiplicity of beings (division of reproduction, schizogenesis, division of protozoans, meiosis in sexed animals, etc.). To be sure, the multiplicity of beings is not absolute; the beings thus reproduced and separated are not worlds totally foreign to one another. The individuals of the same species can be reunited, not only indirectly through the sexual union of gametes, but directly in certain cases (autogamy, fusion of two adjoining eggs, etc.). But they nevertheless escape the dominance of a superior unity. They escape it enough to fight among themselves or to push one another as foreign bodies. Already during the cellular multiplication of development, a certain alterity appears from one cell to the other: the equipotentiality is largely distributed. The "proper body" of a multicellular, of a man, appears to it as *its* body, but all the same as a body, despite the intimacy of possession. In the cellular multiplication of reproduction, alterity is more complete; the individuals of the same species are strangers who can most of the time only touch one another superficially or treat one another as obstacles.

Even identical and Siamese twins fight and beat each other, although an insignificant circumstance probably transformed what should have been a "proper body" into an "other-body." Our A and B (Figure 25) could be a single "Y-shaped monster" who has survived, like the twins who lived until twenty-eight years of age in the court of James IV of Scotland and who had a single common body from the pelvis down.

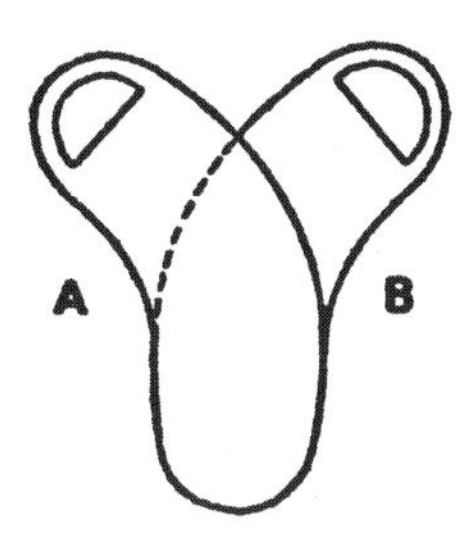

Figure 25.

Through the multiplication of reproduction, the "for-itself" of each after division is enclosed in a kind of impermeable shell and is only interested in others in exceptional cases. Only the "shells" act on one another. The multiplication of reproduction is not a purely biological phenomenon. Because the tiniest bacteria can only be composed of a small number of large molecules,[9] the multiplication of bacteria necessarily looks like a multiplication of molecules (as in catalytic effects) and a true chemical "reproduction."[10] From the molecule to the chemical micelles, from the latter to biological micelles, all sorts of transitions can be found. The chemical or biological micelles are able "to split into similar micelles, by a genuine process of reproductive fission."[11] According to some physicists, the multiplication of beings from a primitive unity is much more fundamental, since G. Lemaître proposed the bold hypothesis of the unique primitive atom, the organ of any cosmogony.[12]

But the interesting point is that multiplicity—and therefore the "body," so-called physical or material existence—emerges from a more primitive unity that is not a body but an autosubjective being, a form for itself. We need not resort to bold cosmogenic hypotheses to uncover, if not to understand, this passage to physical existence. It takes place every day in plain sight. It is not a fabulous adventure that ends in a *Grund*, a dialectical moment of the absolute Spirit, or an "annihilation of the in-itself." When a whole school of herrings emerges from the spawning of a single herring and meanders "mindlessly" in the water like a cloud in the sky, there is indeed a passage from an organic reality to a semiphysical reality.

From all points of view, the mode of being of the body, of physical matter, is a derived and secondary mode. From all points of view, the mode of the subjective, equipotential domain is more fundamental. The "thematic" and the "teleological" are primary. Starting from autosubjective domains, we can understand how a multiplicity of beings can appear through fragmentation or reproduction, a multiplicity that is subjective yet tied by step-by-step connections and whose speed of interaction will constitute what is called physical existence. But the reverse is not true. It would be impossible to understand how a subjective domain could be born from a multiplicity of physical beings that are pure bodies. When a composition seems to be creative, it is because the component bodies have not interacted as bodies but as subjective domains that are not totally distinct.

Cartesian dualism, or the modified dualism we provisionally posited for convenience, can now be abandoned. The intimate union of consciousness and the body, or of the organism as a subjective domain of consciousness and the organism as a set of tools-organs, is not a scandal or an enigma. The body is the appearance that a composite domain B takes for a subjective domain A, when B only acts on A superficially and when A only considers the multiplicity of B's component subindividualities. All kinds of degrees exist in the intimacy of interaction, from the intimate participation of two domains—in which case they are strictly one and are simple parts in the same form in itself—to the nearly absolute distinction that engenders the object as a pure thing. In our own body, we discover all of these degrees, because our nails and hair are for us bodies almost as foreign as a pocketknife or a comb, because we are only interested in their healthy physical state, and because we can see them from the outside and cut them without any cenesthetic sensation. By contrast, our sensory cells directly participate, through their own activity, in the activity of our I-consciousness and contribute to "informing" this consciousness in the two senses of the term.

The general problem of the multiplicity and interaction of beings subsists, but it absorbs as a particular case the problem of the interaction between consciousness and the body. What subsists above all else is the enigma and the paradox of a multiplicity that offers degrees and excludes unity *more or less*. Finalist activity implicates a "surveying" unity that organizes a subordinated and semi-"alienated" multiplicity. The animal or human finalist activity, which uses the brain and sensory organs as devices to make the external bodies intimate and to organize them into tools, is simply one particular case of finalist activity in general. To grasp the general case, we have to examine more closely the nature of the unitary domains of form and activity.

9

"Absolute Surfaces" and Absolute Domains of Survey

Up until now, we have simply opposed the unitary domains of activity (cortical consciousness, embryonic and organic consciousness, individuality of nonstatistical physics) to machines without equipotentiality or to Gestalt-forms with only a pseudo-equipotentiality. Can we define the content of these domains and the relation of their properties to their nature more positively? To begin with, let us consider a simplified case.[1] A physical surface, the surface of a table for instance, can be defined *partes extra partes*. If the surface is checkered (Figure 26), the various fragments of the marquetry will be external to one another.

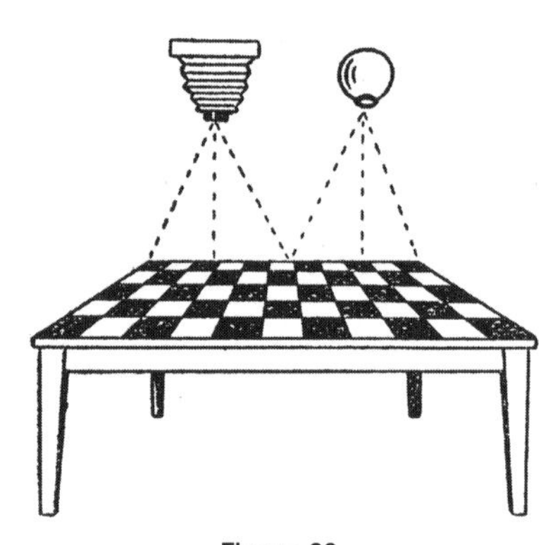

Figure 26.

Relative to any one among them, they are all *somewhere else* on the surface. To capture the entire surface, a camera has to be placed at some distance, along a perpendicular dimension. By the same token, a living being that can be localized as a body must have its eye situated roughly like the camera to perceive the whole surface and its decorative pattern. If I look at a photograph of the table's surface, I will be forced once again to place my eyes at some distance from it. I have to be in a second dimension to photograph or perceive a line. I have to be in a third dimension to photograph or perceive a surface.

We know—it is one of the commonplaces of popular books on mathematics—that one-dimensional beings in a one-dimensional world cannot see a line as a line but only as a point; that infinitely flat beings living on a surface would believe they have sufficiently protected a treasure T by enclosing it within a circle that deters the indigenous thieves V, V', V"; but a thief evolving like us in a third dimension could touch T without touching the protective circle (Figure 27). By analogy, it is easy to conclude that all the points of our solid bodies are simultaneously visible to an observer who exists in a fourth dimension. Solid bodies are "open" in the fourth dimension as a circle is open in the third. A four-dimensional being could see and pierce our heart without touching our skin. In short, an observer always has to be situated in the $n + 1$ dimension to see at once all the component points of an n-dimensional being. And yet *this geometric law, which applies to the technique of perception, that is, to perception as a physicophysiological event, is invalid for visual sensation as a state of consciousness.*

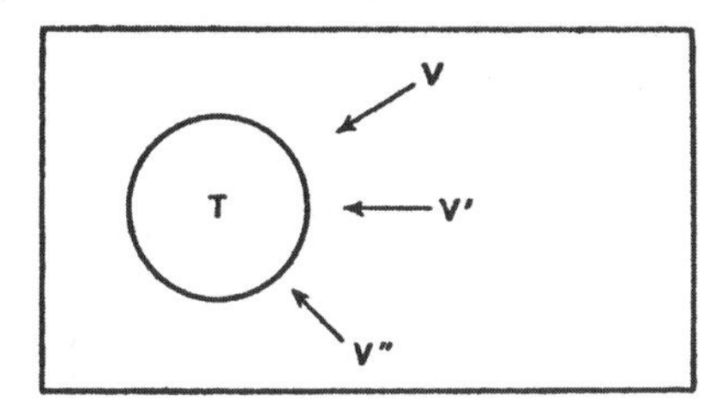

Figure 27.

Let us shift our attention from photographic observation and the organic mise-en-scène of perception to my visual sensation in itself. Like the table or the photograph of the table, it contains multiple details, checkers which are also in a sense *partes extra partes,* each existing at a different location from the others. This time, however, "I" do not need to be outside my sensation, in a perpendicular dimension, to consider each and all the details of this sensation. Even when, instead of fixing my attention on the table, I "inspect" my sensation (to register my astigmatism or my myopia), I do not have to place myself outside my sensation to know it. If I were to observe the cortex of a being in the process of looking at the table, I would have to be outside this cortex; but if I were experiencing my own sensation, I would not have to separate myself from it. This is fortunate fact, because I would otherwise need a third eye to see what my first two eyes see, then a fourth to see what my third sees, and so on.

I would be like the man J. W. Dunne[2] speaks of, who, wanting to create a complete painting of the universe, (1) first paints the landscape, (2) then realizes that he forgot himself and represents himself in the act of painting, (3) then realizes that he forgot to represent himself in the act of painting himself, and so forth (Figure 28). Because it is consciousness-knowledge and not observation-knowledge, self-enjoyment essentially dispenses with infinite regress and a "serial universe." Dunne believes that infinite regress is inevitable because he turns knowledge and consciousness into a kind of observation or, as he says, "description." The observation of an experience must then be, once again, the observation and description of this experience as my own. But another observer has to observe and describe the second observer, who observes and describes the first, and so on. In fact, as Dunne says, "the mind which any science can describe can never be an adequate representation of the mind which can make that science."[3] From this perfectly true thesis, Dunne draws a perfectly false conclusion: "the process of correcting that inadequacy must follow the serial steps of an infinite regress."[4]

Obviously the right solution is that the "description" or "observation" of the mind (or the subjective domain) is a whole other matter than the subjectivity of the described or observed "mind."

Dunne's conception, though it amused many people, has not had great success in contemporary philosophy. But perhaps we have not carefully examined the consequences of the negation of infinite regress. Let us return to the surface of the seen-table. It does not obey geometric laws. It is a surface seized in all of its details, without a third dimension. It is an "absolute surface," which is not relative to any point of view external to it, which knows itself without observing itself. If I were to place my eye on the table, I would see nothing, but I need not be "at a distance" from the sensation to see it extended. In contrast, I cannot turn around the sensation to consider it from various angles. "I" (my organism) can turn around the table to *obtain* different sensations, but "I" cannot turn around my sensation once I obtain it.

Figure 28.

The seen-table is also a one-sided surface (like the Möbius surface, but in an entirely different sense): if I saw in my visual field a peripheral luminous stain moving forward, no mental procedure would allow me to see it moving backward (as an oculist who looks at my retina would observe it in his ophthalmoscope). This fact is tied to the nongeometric nature of conscious survey. If the perceptible surface could be seen from both sides, it would not be a sensation but an object.

As experience demonstrates, I can turn my attention or my "mental prospection" to this or that detail of the sensation without moving my eyes—for instance, to this white or black square. I can swap the black or white squares in their roles as figure and ground, but these "displacements" of the internal observation do not obey the laws of physical displacement and observation and do not have the same effects. The sensation's multiple details are distinct from one another, and yet they are not truly *other* for one another, because they constitute my unified sensation. They have a well-determined order; they even have metrical relations (e.g., the squares appear equal), but this order or equality does not have a purely operational value, like the technique of the craftsman who inlaid the table. Order and multiple relations are immediately given in an absolute unity, which is nevertheless not a fusion or a confusion. This amounts to saying that my sensation is a form proper, a form and not a pattern, a structure, an assemblage of elements, or a Gestalt-form.

Relative to the multiplicity of details in my sensation, "I"—the indefinable "I"—appears as the unity, as a unity endowed with ubiquity. Here as well, sensation and subjectivity generally escape the ordinary laws of physics. It has been said that the core of the theory of (special) relativity amounts to the realization that one cannot be in two locations at once. In this sense, the absolute subjective expanse escapes the jurisdiction of the theory of relativity. "I" am simultaneously in all the locations of my visual field. There is no step-by-step propagation, no limit speed, for such a domain. If I look at two clocks in a single glimpse, they will be one, despite their difference. There is no "absolute elsewhere" in a subjective domain, because there is no absolute alterity between details. If I were to number the cases of the checkerboard, the squares at one end would be farther away from the squares at the opposite end than from the middle squares. And yet this variable distance, which appears in the ordered figure of sensation, is not a true distance that would require physical means and energy to be overcome.

The notion of absolute survey, of nondimensional survey, is the key not only to the problem of consciousness but also to the problem of life. It allows us to grasp the difference between primary consciousness and secondary consciousness, a problem we have already tackled.[5] Since the question is a difficult one, let us reflect on concrete cases with the help of images.

a. To begin with, let us schematize a man writing on a cluttered table as seen by an observer and, on the other hand, a protozoan (the example of a living being with a nervous system) in the process of skirting an obstacle by trial and error (Figures 29 and 30). The observer sees the man turn his head and eyes, that is, his attention, toward the objects placed on the table. He can measure the distance between the man's eyes and his paper as well as the distance between the protozoan and the obstacle. Similarly, he can follow the progression of optical stimuli and neural influxes from the seen objects to the retina, to the occipital area, to the motor cortical centers, and then to the medullary centers.

b. Let us now suppose that I myself am the seated man. Here is what my visual field affords me (Figure 31). This visual field immediately presents both my body (of my head, only the vague circle of my glasses and the more vague images of my nose and lips are visible) and

Figure 29.

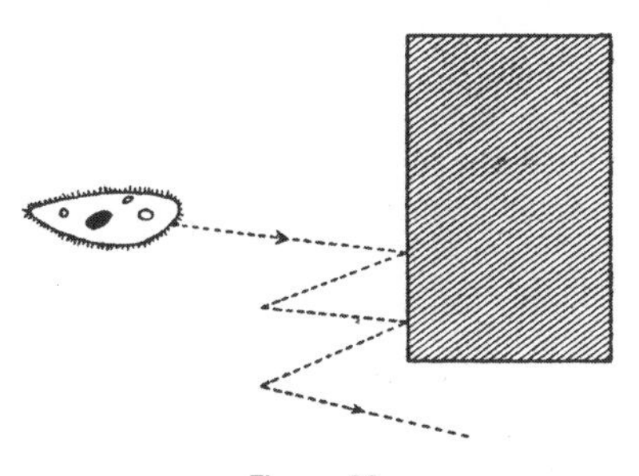

Figure 30.

the objects I observe, namely, my table, the books that clutter it, and the paper on which I am writing. A perceptible distance immediately appears between my seen-body and the seen-table, a distance that seems to correspond to the distance between my real body and the real table that the observer is measuring.

Biology teaches us that this field of sensory consciousness is localized in my occipital cortex; it is probably the very reality of my area striata or of a certain level of this area.[6] But at any rate what is certain is that all the details of the sensory image have to be given immediately in an absolute unity, because there is no third retina or second

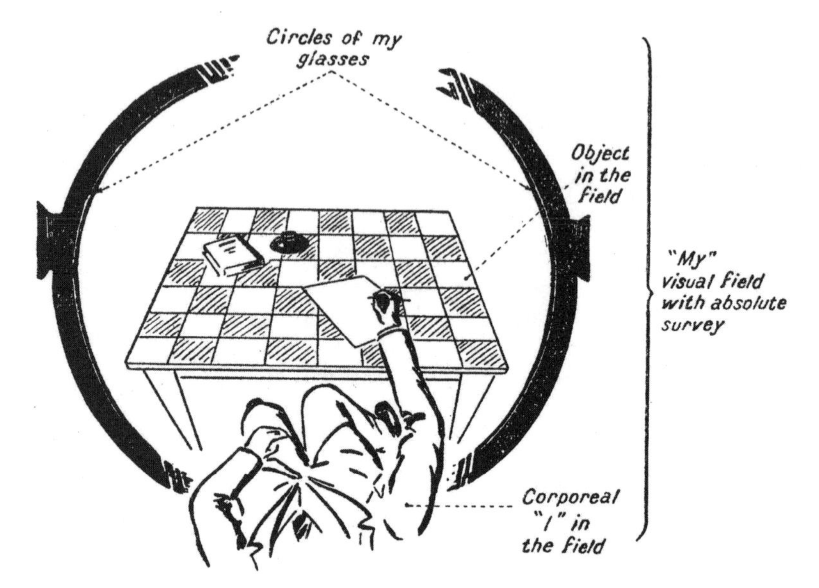

Figure 31.

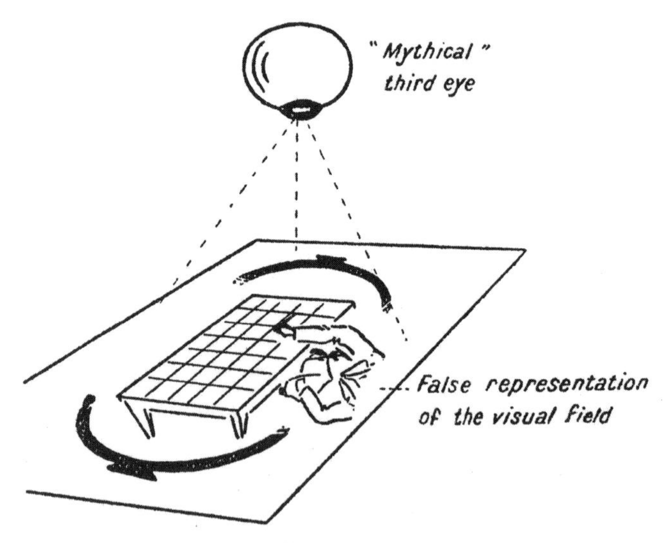

Figure 32.

striatal area that can see this visual field from the outside as the observer sees the man in the process of writing. The "I" or the conscious unity, whatever it may be, has the acute impression of surveying this field of consciousness as though it observed it from the outside. I can hardly resist the temptation to imagine myself, to imagine the "I," above the apparent circle of my glasses, by identifying this "I-unity" with a kind of center of the invisible head that my sensation allows me to presuppose. And yet it is clear that the "I," the unity of consciousness, is not at a distance, in a perpendicular dimension, from the totality of the visual field in the same way that my eyes and my head are at a distance from the paper on which my hand is writing. The image of my glasses and the vague shadow of my nose and my eyebrows form part of my visual field. Thus the biologist who observes me from the outside can localize all of these perceptible forms, like all the images of my body, in my area striata, where—it is worth repeating—there is no third eye. My visual field necessarily sees itself through an "absolute" or "nondimensional survey." It surveys itself without positioning itself at a distance and in a perpendicular dimension.

It is therefore a gross error to imagine the visual field in the occipital area as a kind of photograph, or as those cinematographic montages in which a three-dimensional scene suddenly becomes an album page that begins to turn before us on the screen. Between the "I-unity" and the visual field, there is only a purely symbolic "distance" (Figure 32).

Assuming we accept the natural hypothesis that the visual field has some connection to the occipital area, the visual sensation proves then that at least a certain part of the organism is capable of direct self-consciousness: it sees itself through absolute survey, without any observer in a perpendicular dimension.

c. Because the occipital area, which is modulated by optical stimuli, ultimately has to see itself, to enjoy itself, why couldn't the protozoan "see" itself directly just as much as our cortical tissue? The protozoan has neither eyes nor mirror; but neither does our cortex have an eye or a mirror to see what the eyes have already brought it. Seeing itself, the protozoan or its "unity" in absolute survey will not see external forms in this field of self-enjoyment (it will not see, for instance, the form of the obstacle it is trying to skirt). It has no sensory organs that would permit the modulation of a part of its organism according to the pattern of external objects. Its field of consciousness will only be its own organic form, which is in principle the entire universe for it. This surveyed, organic form could be as distinct as our visual field and could present all the structural details of the cytoplasmic architecture as clearly as our visual sensation presents all the details of the checkered and cluttered table we are looking at. This organic form or primary consciousness is not vague or psychoid. It has no reason to be so. It can never even be "myopic for itself," like a visual sensation in the secondary consciousness, because it is not our occipital cortex that is myopic but our eyes.

In other words, there is at bottom only a single mode of consciousness: primary consciousness, form-in-itself of every organism and at one with life. The secondary, sensory consciousness is the primary consciousness of cerebral areas. Because the cortex is modulated by external stimuli, sensory consciousness gives us the form of external objects. But this particular content does not represent an essential trait of consciousness and life. There is no reason to deny subjectivity, primary consciousness, self-survey, and the self-enjoyment of their own form to our noncortical and even nonneural cells or to our organism in general. The "I" does not participate in this self-enjoyment because it is specialized in sensory consciousness.

It is not surprising that the "I" of secondary consciousness should be irremediably cut off from primary consciousness, that "I" should have no direct primary consciousness of my organism. This disconnection represents a normal phenomenon of "distribution," like the "distributions" that fragment the areas of development in the course of embryogenesis and "determine" them by specializing them. Cenesthesia, as we have seen,[7] has nothing to do with primary consciousness. It is a secondary consciousness in the same way as visual consciousness; both presuppose a healthy cortical area (parietal area). Likewise, the instinctive drives and the sensations of organic need, which emerge in the secondary consciousness, cannot give the "I-consciousness" any intuition of the essence of primary consciousness. To believe that they do is an inexhaustible source of philosophical error, for by imagining organic consciousness on the model of the drives through which it communicates with the secondary consciousness, we attribute to it, for no good reason, the vague and confused character that belongs uniquely to these messengers.

Contrary to an ingrained prejudice, consciousness or the *x* unity of nondimensional survey is not essentially perceptive or cognitive of spatiotemporal structures. It is essentially active and dynamic; it organizes spatiotemporal (organic or sensory) structures that are given to it in its field of survey. Consciousness cognizes only ideas-forms, themes, or transspatial types, at which it aims beyond the field of survey and according to which, as ideals or norms, it organizes or improves the organization of structures-forms in the field.

This is the most delicate point of our difficult question. We should vehemently deny the existence of a geometric dimension that provides a point of observation external to the sensory field. But we should affirm no less vehemently the existence of a sort of "metaphysical" transversal to the entire field, whose two "extremities" are the "I" (or the *x* of organic individuality), on one hand, and the guiding Idea of organization, on the other.

For the primary consciousness (e.g., the protozoan's), the guiding Ideal is the organic type. For the secondary consciousness of an animal with a nervous system and sensory organs, the guiding Ideal is both the organic type and an *Umwelt* intimately connected to this type, according to which the bee, for instance, only sees in the external forms captured by its sensory organs the flowers as reserves of nourishment, the hive as refuge, and so on, and searches for and maintains them in this state. For the human secondary consciousness, the guiding Ideal is the world of essences and values, detached from the human organic type. But in these three cases, consciousness is not an inert domain that is simply unified by the absolute survey; consciousness is organizing. The protozoan strives to maintain its organic type despite the physicochemical phenomena that tend to alter it. The bee shapes the world according to the instinctive gnosia that characterize its specific *Umwelt.* "I" strive, for example (Figure 33), to tidy up my seen-table by referring to an ideal of order; or I strive to maintain my tools in good condition; or, more generally, I strive to realize my ideal norms by incarnating them in the beings and objects that surround me.

Up to now, we have proceeded as if "absolute domain" were synonymous with "absolute surface" and our schemas have accentuated this impression. But because the absolute surface is intuited without a third dimension, nothing in fact prevents us from conceiving more

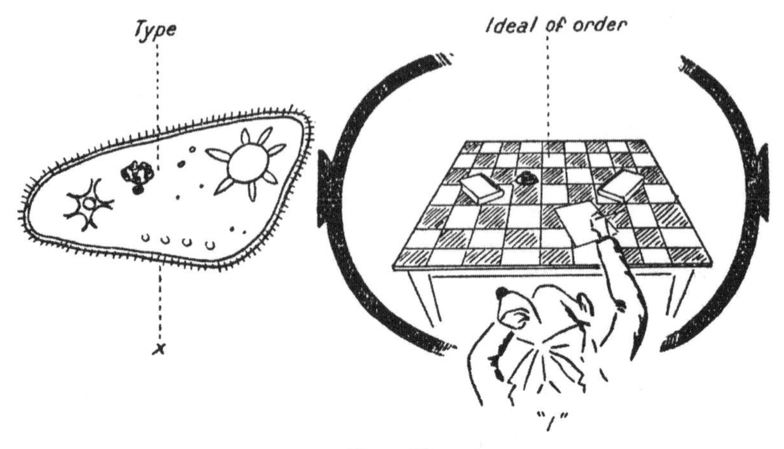

Figure 33.

general absolute domains—absolute volumes, for example. Primary organic consciousness has to resemble an absolute volume rather than an absolute surface because, when observed as a body, it appears as a volume. But because the geometric laws do not apply to the subjective domains, the primary consciousness of a three-dimensional organism, while constituting a form in which all the details are simultaneously present, does not require the hypothesis of a subject lodged in a fourth dimension. Primary organic consciousness must even correspond to an absolute domain of space-time. The organism is never an instantaneous anatomical structure; it is, rather, a cluster of processes. A species is characterized as much by the phases of its development as by its adult form. A "type" is spatiotemporal. Its embryological forms are part of its anatomy in space-time; its development is inseparable from its being. In principle, absolute domains imply a possibility of time-survey and space-survey, but with limitations to which we will return. In the case of absolute domains, it is the whole space-time of physicists that has to be "surveyed" without any supplementary dimension.[8]

The survey of the "I" is purely metaphorical. The "inspection" that the "I" seems to perform on its domain of survey is equally metaphorical. In fact, domain, "I," and Ideal form an indissociable whole that *is* active inspection; a different "inspection" corresponds to a change in the domain, a change in the "figure," or a figure–ground mutation. The role of the subjective domain in the regulation of subordinated organic mechanisms and tools is thus clear. These organic tools are not pure tools that are simply inspected by a custodian or a worker in the flesh. The extraorganic material tools and factories elude in great part their proprietor. Humans cannot be everywhere at once to ensure that everything runs smoothly and to repair what deteriorates. By contrast, the organic tools, at least in young organisms, are "maintained" by subjective equipotential domains that "survey" and "inspect" them with the ubiquity inherent to subjective domains and to absolute surfaces, that repair them in case of light wear or lesion by correcting the blind operation of subordinate amboceptors.

The difference between the inspector in the flesh (relative to his extraorganic tools) and the field of inspection of organs is the same as the difference between the physical and technical conditions of observation and those of conscious sensation. In both cases, it is necessary to come to a stop without ascending to infinity. If a tiny internal inspector had to oversee the organism of the engineer by wandering in him as the engineer wanders in the factory, who would oversee this internal inspector? Very fortunately for us, the inspection of our organs is final and absolute; it is self-inspection. By keeping track in his office of drawings and graphs that reproduce the state of machines and supplies in the distant factory, the engineer tries to imitate the mode of organic and cortical inspection. These graphs and drawings can be seen all at once, while the real factory operates semiblindly, by a succession of productions and services. And the engineer can avert a lack of coordination that emerges in the graphs before it is really experienced in the services. This "artificial cortex" must nonetheless rely on the real cortex of the engineer, who is an absolute surface, a drawing that reads itself.

Here again, there is no doubt that absolute surfaces and absolute autosubjective domains are primary relative to all the categories of pseudo-forms, *patterns, structures,* various *assemblages, Gestalten,* and so forth, and cannot be composed of them. The drawings and graphs in the engineer's office postdate the factory, just as the visual sensation of the checkered table postdates to the table. But the engineer who built and assembled the factory clearly had "in mind" an overall outline of this assemblage, just as the craftsman who created the checkerboard "saw" it or referred to its image.

If absolute surfaces are accepted as primary, then another paradox will arise. The history of evolution seems to require the idea of a progressive formation instead of preexisting absolute forms. The engineer's cortex (as well as his consciousness) precedes the drawings and graphs he uses; it was formed in the course of embryogenic development. But, as we have seen, this cortex simply retains the equipotentiality of the embryo that derives from the equipotentiality of the egg, which in its turn derives from the germinal equipotential cells. Because equipotentiality is the typical manifestation of absolute forms, it can be said that as high as one climbs in the history of living forms, one always discovers an absolute form that has subsisted uninterruptedly for hundreds of millions of years of biological evolution. From primitive living beings to humans and their brains, formation does take place, but this formation starts from a different absolute form and not from dispersed elements. There is a formation by continuous improvement in the constant presence of an organic domain. It is never a question of formation through the assemblage of bits and scraps.

If there is, strictly speaking, no beginning for absolute domains, there cannot in principle be any end. In fact, we do not see how a subjective domain of self-inspection could come to an end on its own. Aging and death are conceivable only in the case of a secondary inspection (like the engineer's inspection of a factory) bearing on machinery that is itself detached from organic subjectivity and repaired only at long intervals. The body of a metazoan is made up of organs that, macroscopically, are quasi-autonomous factories subject to the risk of equally macroscopic accidents. The possibility of replacing these organs with automata is the underside of their perishability. In contrast, the impossibility of replacing the living tissues as such with constructed automata is the underside of their imperishability. There are indeed microorgans in a protozoan, in a germinal cell, or in the cells of a tissue cultivated in vitro. But we should realize that these microorgans are not made up of autonomous amboceptors; that subjective "inspection" is total and perfect, because all these living beings are potentially immortal; and that, from germ to germ or from cell to cell, none of the currently living cells, derived by division or fusion from other cells, has ever died. The heart, as a large innervated and irrigated muscle, can malfunction, but the cardiac tissue with its embryonic rhythm is theoretically immortal.

There is certainly a relationship between immortality and equipotentiality, because equipotentiality enables the regulation of lesions, because Lashley could not have carried out on the rat's heart the interventions he made on its cortex, and because embryologists can slice an egg or a young *Triton gastrula* in two without killing it, whereas a sagittal or other cut of an adult *T. gastrula* would infallibly kill it. Like equipotentiality, virtual immortality is the sign of the presence of an absolute domain, whose primary inspection maintains its form indefinitely. It is the sign that the microorgans' order of magnitude is related to the order of magnitude of the dynamism inherent in primary subjective bonds. If virtual immortality is rarely real, it is because even an absolute domain can be violently destroyed by relatively immense forces, which result from accumulation in the world of physical aggregates. Even though its bonds may have a primary order relative to the step-by-step bonds of the physical world, they are quantitatively too weak to resist these forces. Owing to their more accentuated unity, the absolute domains of physics (atomic or subatomic individualities) have by contrast considerable binding energies. They are virtually immortal. It is well known that the disintegration of an atom is quite a story, much more so than the disintegration of a human being.

10

Absolute Domains and Bonds

As we just saw, the absolute domains of survey are not assemblages of bits and scraps. Even though it is complex and supplies the details of the pattern, the sensation of the surface is not composed of small squares glued together, just as they are on the physical table. And yet, we also spoke of the internal bonds [*liaison*] of an absolute domain and even of the variable energy of these bonds. A bond implies, it seems, bound parts.

There is no contradiction here. The internal bonds do not explain the domains; on the contrary, it is the domains that explain the bonds. The analysis of the notion of "bond" shows that it implies an absolute domain of self-survey. This is even one of the shortest paths by which to arrive at the idea of "absolute survey."

In passing through the history of scientific philosophy, we notice that a concept as significant as the "bond" was greatly neglected. It is true that science has not tackled the problem in earnest until quite recently, with wave mechanics and Heitler's and London's research on molecular bonds. Previously, philosophers had to confine themselves to notions drawn from global physical experience, such as "plenitude," "solidity" (in the Democritean sense), "attraction," and "field," or to engage in abstract discussions on internal relations and external relations à la Bradley and his admirers or opponents. By definition, pure observation cannot reveal the bonds of the observed being, because the waves or photons it emits only retain a *pattern* of it, without internal bonding or with bonds that differ completely from those of the observed being. We can only arrive at the proper bonds of an object in two ways: abstractly, by an induction drawn from prolonged observations, which shows the degree and mode of consistency of the object, or concretely, by a kind of analogical animation that seizes the being behind the object and "knows" its proper sense or coherence. Let us consider, for example, the glue that binds the fragments of the marquetry. How can this glue

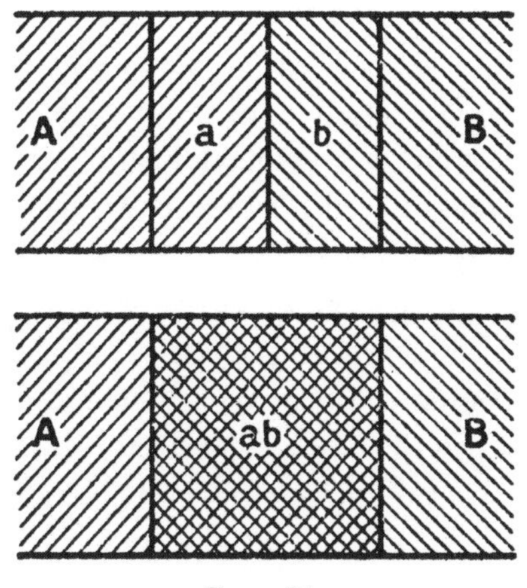

Figure 34.

bind? Beyond the explanations that derive from science's "average regions" (properties of colloidal micelles, structure of molecules), we have to reach the moment when physical elements, contiguous to one another, are nevertheless immediately unified. So we encounter here again either the necessity of an infinite regress (a glue is needed between the elements of the glue for it to be adhesive and so forth) or a domain of absolute bonds. This domain of bonds is none other than the domain of absolute survey we already know. Let *a* and *b* be bound elements (that we can assume to be "binders" of other elements A and B) (Figure 34). If they are simply juxtaposed, both as observed objects and in their very being, each absolutely enclosed on itself and each really "next" to the other, how can they be unified and how can they serve to unify A and B? But if their domains are superimposed on one another, and if the superimposition is not understood once more as a simple juxtaposition or a simple mixture—which would not clarify the matter, because the mixture would bring us back to the "average explanations" of science and would implicate subindividualities of *a* and *b*, α and β, which are themselves juxtaposed—if the domain of superimposition is at once *a* and *b*, if it is *ab* considered as a new being with an autosubjectivity and a self-survey, then the bond can be understood.

Contemporary physics discovered that the interaction of similar particles bound in a system is necessarily correlated to a loss of the individuality of these particles. Variables *a* and *b* can no longer be identified in an absolute way; they no longer represent impenetrable domains in space; they are domains of the possible localization of what we observe as a corpuscle. In the domain *ab,* which belongs both to *a* and to *b, a* and *b* become indiscernible; and it is no longer possible at any moment to tell whether we are dealing with *a* or with *b.* Thus even the domains that are schematized as purely *a* or purely *b* can no longer be so, because in the mixed domains, *a* and *b* exchange their roles. This possibility of role exchange in the course of *a* and *b*'s interaction in the common domain *ab* is dynamically translated as exchange energy, and it is the basis of chemical valence or, more exactly, of covalence, in contrast to the heteropolar bond between ions (sodium chloride type).

No doubt the spatial schema (on the physical surface of this page) completely distorts the notion of bonding: *ab* appears as *partes extra partes.* The spatial schema also distorts the physical interaction, which cannot be represented in the ordinary three-dimensional space. But the essential point is that the binding energy appears at the moment when the elements bound within a system lose their individuality. The domain *a,* the domain *b,* and the domain *ab* cannot be understood as spatial in the ordinary sense of the term; rather, they are absolute domains like the sensation-table, which is not "spatial" but a form, a unitary system.

In this sense, it can be said that a field of consciousness or of subjectivity is a typical domain of bonds, on the model of which we should conceive the domains of microscopic bonds that ensure the coherence of physical individualities and, indirectly, the solidity of physical aggregates, through step-by-step interassemblage. Pascal Jordan proposes to consider even the "largest" organisms as part of the world of microphysics, because they have an organic unity and a unity of behavior that can be linked (through the genes) to systems on the order of magnitude of atomic systems.[1] Provided we do not insist too much on questionable considerations,[2] this idea is perfectly true. The elephant, we might say, is a macro-microscopic being. Likewise, and by taking things from the other end, we can consider the bonds inherent to the absolute domains to be of the same general type as the bonds of microphysics. A field of consciousness seems to be too "vast"—if this term is meaningful—and complex to represent the schematic type of bonding, in the same way that a large mammal cannot be easily described as "microscopic." And yet, like the elephant's organism, human consciousness has a more primary unity, a more primary type of bonding, than a grain of sand. The primary type of every bond is "absolute survey," that is, being-together as immediate form. The glue can glue, and steel or diamond can be solid, only through the microscopic action of domains of absolute survey. We invert things when we explain the unity of an equipotential domain through bonds or fields borrowed from the order of a macroscopic physics, which has only retained the step-by-step action from the phenomenon and not the elementary bonds that can make the "step-by-step" binding and the glue adhesive. Physiology and philosophy insist almost exclusively on consciousness as knowledge; consciousness is also a binding force.

Consciousness is indissociably both knowledge and binding force. Let us imagine that only a single man remains in the universe, ending his life as a hermit, repairing his cabin, cultivating his garden, and creating some tools. Without dying, he becomes unconscious. The suppression of his consciousness will not be ineffective, as the epiphenomenists claim. Nor will it be totally effective: the flowers will continue to grow in the garden; the cabin will not collapse immediately. The suppression of consciousness will nevertheless condemn this small human world to eventual dissolution. The unmaintained cabin will fall into ruins; the garden will become wild again. In short, the suppression of conscious connections will suppress the corresponding forms. Consciousness is cognitive relative to ideals and "binding" relative to physical beings, which it informs according to these ideals.

We discover similar results by following another order of considerations. As a unity in the multiplicity, an absolute domain or a true form realizes the otherwise inconceivable synthesis of being and having. *Is* the system *ab a* and *b*, or does it *have a* and *b* as parts? Does the surveying unity have the details it surveys, or, because the survey is purely metaphorical, is it the very totality of the surveyed details? The term *to be* signifies in this case "to consist of"; having is opposed to being only in this sense. If "being" is taken in the sense of "proper existence," having will in contrast presuppose being. Ultimately, if *to be* signifies "to consist of," the subject of the verb *to be* is only a linguistic convenience, because it simply designates the whole of its component elements. Because it is nothing by itself, it can possess nothing. A manufactured machine or a piece of furniture like the inlaid table does not float as a unity above its components. If the word *table* is employed as the subject of a predicative phrase, this would be a simple manner of speaking and an implicit reference to the machine or to the piece of furniture conceived by the engineer or the craftsman. In effect, the table was drawn as a table, that is, as the theme of a table, with determined traits that the craftsman selected. It is from domains of subjectivity that material objects borrow their unitary being and the possibility of "having" their properties. The absolute domains must therefore engender, on their own, the synthesis of the unity of being in the multiplicity of havings.

Is a water molecule (does it consist of) two hydrogen atoms and one oxygen atom, or does it have, as its unity and being, three component atoms? We see straight away that this problem has the same solution as the problem of the bonding of elements and the partial loss of their individuality in the unity of the interacting system. According to wave mechanics (Figure 35), in the schema of the water molecule, the wave functions of the three atoms partially overlap. Then an interaction energy appears. But this partial overlapping implies a partial loss of the individuality of the electrons involved in the valences. This loss is gained by the molecular system, which is thus a genuine unity and, in this sense, "possesses" the three atoms. If there were no zone of overlapping, the molecule would only *consist* of three atoms—or rather, there would be no molecule at all.

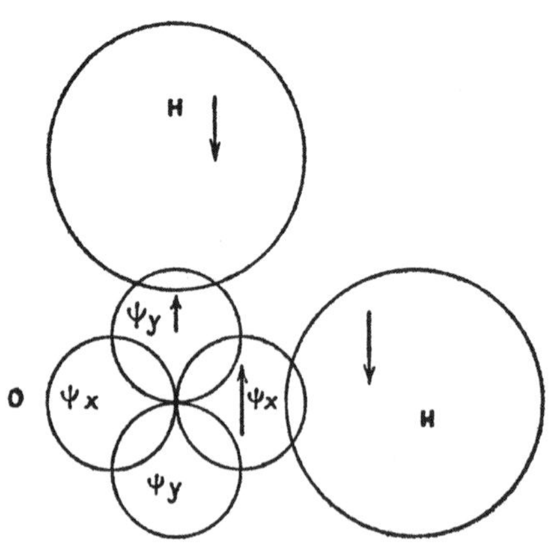

Figure 35.

Considered in their being and not merely observed, organic domains present a particularly typical case. In the course of mitosis (e.g., at the moment of the constitution of a spindle between two attractive spheres), does the cell have two poles of attraction, or is it already two individualities? The passage from unity to duality has to be progressive, like the passage from prophase to telophase. Obviously, if the cell were merely a geometric objective domain *partes extra partes,* this progressivity would be inconceivable. The cell has to be an absolute form with self-survey to control the beginning of its own division, progressively diminishing the unity of the system for the benefit of the individuality of its components. If it is a matter of a mitosis of development, the system's unity will not disappear completely, because a unique theme can be distributed in two daughter cells (which would become, for instance, the right half and left half of the same organism). If it is a matter of a division of reproduction, the unity of the system will disappear completely. But this is perhaps only a semblance, because the two individuals of the same species can eventually form a colony. If it is a matter of two identical twins, they might sometimes be mirrored as though they were at once two individuals and the two halves, right and left, of a single individual. One "has" a right half and a left half, but "one" eventually could "have been" two individuals or *nearly* two individuals, as in the limit cases of division where the duality is discretely marked only by the splitting of the nose and the rudiment of a third eye (Figure 36).[3] The passage from a domain of absolute survey to two domains—even if they are presumed to be still connected to a specific unity—is of course enigmatic. But if we do not posit the notion of "domain of absolute survey," the "having" or the "being that has properties" (in contrast to the being that merely "consists of") will be nothing more than an empty word or, if we wanted to realize it, a contradiction in terms.

Figure 36.

The metaphor of "possession" is instructive. If we consider the objective aspect of the fact, then the possession of a table, of a machine, or of a house will only designate a conventional series of acts, even independently of the juridical convention of property. Psychologically, the fact that the table, the machine, or the house appears as a familiar sensation in consciousness gives the possession an absolute and immediate reality, and thereby the possession always modifies the possessor in its being. The "proprietor," or, as J. Galsworthy says, "the man of property,"[4] is a well-known type of man. To have a visual sensation is simultaneously to be. The sensory cells' individual activity is not lost in an ensuing global and massive unity, because the details of my sensation depend on this individuality and remain distinct in the surveying unity of the absolute surface. "I" possess this sensory activity, and my possession totally transcends the possession of an object through an external relation. I participate in it, I am modified by it, while remaining distinct as a metaphorically surveying unity. "Being-having" simply designates the domain of survey and the mode of bonding of parts in an absolute form. It would be obviously absurd to imagine that a molecule's mode of unity is the same as an organism's and that the fusion of primordia in the case of the accidental abortion of the median embryonic primordium is the same phenomenon as the bond of formation of homopolar molecules. The differences are manifest. But this is all to say that at the root of these two problems lies a common given. The various mysteries we have encountered converge in the primary mystery of the form-in-itself.

11

Absolute Domains and Finality

The general traits of the absolute domains we have examined are the conditions of finalist activity, but they are not finalist activity itself. By contrast, the traits we will now study blend with finalist activity as we have described it.

a. *Thematic forms.* In an absolute domain, there are no stereotyped patterns; there are instead true forms, which, as it were, follow and recognize one another in their various aspects. Already in organic development, as we have seen, a muscle "knows its name," according to P. Weiss's picturesque theory. In the order of instinct, the forms that concern gnosia and praxia in the *Umwelt* are "recognized"; they do not function mechanically like keys in a latch. The experiments of animal psychology have also shown that for the chimpanzee, a baton, for example, is not an opticogeometric pattern *ne varietur* but "any prehensile elongated object." In a behavior guided by the consciousness of values rather than by needs or instincts, the forms are also recognized according to their relations with the intended values, or inversely, the values are recognized through the forms. If I test various keys on a latch to identify the "right" one, the value will not be tied to the form proper but to the pattern of the key. This limit case is obviously just a degraded case. The absolute forms directly disclose their various values in all the orders (technical, theoretical, aesthetic, etc.) without a point-by-point confrontation between two structures. To use the Platonic metaphor, an absolute domain is a kind of mirror with reminiscence, in which a transspatial projects and recognizes itself. An absolute form is at once structure and idea, ειδος in the double sense of the term. The triangle, for instance, is at once spatial and "ideal." A triangular pattern, serving as the key to a photoelectric screen, has to be strictly defined in space. The same is not true for a triangle. If it is a figure on the ground of an absolute domain, then insofar as it remains within the limits of its "ideal" definition, it will always be "recognized" as a triangle, regardless of its purely spatial pattern. To use more modern terms (but terms essentially equivalent to the Platonic language), the absolute domains enable an "eidetic analysis" and make it possible to discern the world of values and of essences through the spatiotemporal world.

b. *The possible and the necessary.* Provisionally substantializing the "I"-unity, we can say that the absolute domain allows "modality" within behavior, that is, behavior with the sense of the possible and the necessary. We can easily grasp the intimate relation between the immediate possession of a thematic form that has a sense and the behavior according to the possible or the necessary. Consider three itineraries (1, 2, 3) between A and B (Figure 37). The three itineraries are different as brute facts, but they are "equivalent" (with identical value and sense) insofar as they equally arrive at point B, considered as the goal, and are equally possible. As "step-by-step" trajectories, they are absolutely different, just as the different triangular patterns are absolutely different. They are equivalent only in the absolute survey that simultaneously "sees" A, B, and the three trajectories between an infinity of other possible and virtual trajectories starting from A and arriving at B. Let us consider another example (Figure 38). I want to cut out capital letters from a piece of cardboard. For *I*, *N*, and *M*, this is straightforward; but I immediately see, without needing to try, that it is impossible to cut out *O*, *A*, or *B* in the same way and that it is *necessary* to leave bridges to hold the inner surface of these letters.

c. *Temporal survey and finality.* At the level of the secondary consciousness, there is a dissymmetry between the spatial survey and the temporal survey in absolute domains. It is perhaps a matter of a temporary limitation, like the one that barred for so long the access of abstract intelligence to animal consciousness. It is perhaps significant that the authors of great utopian works of science fiction, like Renan, Haldane, and Stapledon, all imagine future humans as rulers of time as well as space.

In O. Stapledon's great science fiction novel (*Last and First Men*, 1930), the last men, having migrated to Neptune, become masters not only of interstellar space but also of time. They can act telepathically on the past; they can guide and rescue it, and this is how—our author ingeniously imagines—a Neptunian who will live some millions of years from now is telepathically dictating the tale to him. In the crudely materialist form of telepathy, this play with time is naturally absurd.

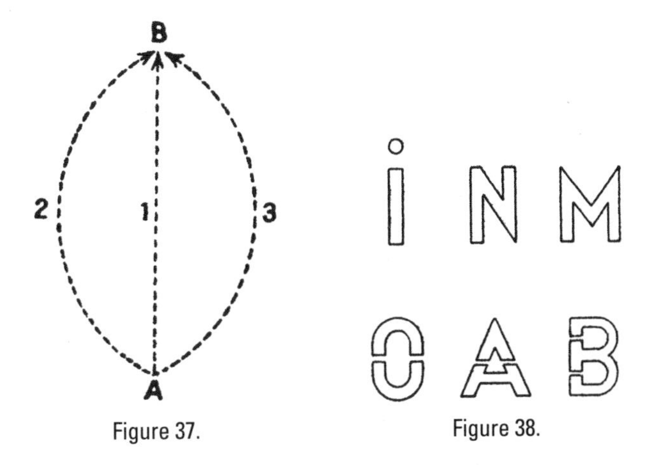

Figure 37.

Figure 38.

The Neptunian in question does not saw the branch on which he is seated; he does better: he sows the grain that will become the tree on whose branch he is seated.

And yet it is this impossibility that is realized in knowledge-history, not materially or psychically but mentally. As R. Aron emphasized, there is a recurrence of the present on the past, certainly not through a causal, material, or psychic step-by-step influence traveling backward but by virtue of the ubiquity of sense.[1] The incidents of the present retrospectively confer a different or variable sense on the incidents of the past. "Is the French Revolution finished?" Cournot wondered.[2] Depending on whether it is finished, its sense is different or could become different. Because it is up to us to extend the Revolution, its sense—and hence its historical being—depends on us. When Hitler believed himself triumphant, he said, addressing the dead Germans of Verdun, "You fell on the path of great Germany," and at that precise moment, if he could have stabilized his tentative victory, he would have been right. Once he was vanquished, the deaths became useless again. These historical fluctuations neither warm nor cool the bones that are turning white, but they indisputably alter the sense of Germany's whole previous history. The recurrence of "sense" in history is tied to the fact that humanity has a continuous life, which surpasses the life of individuals, and that in a precarious yet certain sense, this life is an absolute domain of temporal survey.

For the individual, the mental survey of time is limited to the duration of his life. But within this limit, ubiquity is realized all the more clearly. Humans can at the last moment ruin or save a long friendship or a long fidelity; they can compensate for a long negligence. If this last moment is just a physiological or psychic crisis, there is no virtue in its recurrence. It must have a mental sense so that it alters the meaning of life as a whole by surveying it. In reality, religions are not mistaken in their belief that a final repentance effaces all sins, although simple minds like the emperor Constantine are wrong to imagine that a cunningly delayed baptism can exert a magical action and expunge all crimes. Whereas in the order of space, even at the level of perception and psychic thematism, the sense of the surveyed form is immediately everywhere, a difficult symbolic assembly and a mental culture are needed in the order of time for sense to attain a relative temporal ubiquity. The purely psychic, spontaneous temporal amplitude is not null, even in animals: Pavlov's dogs can be conditioned by the rhythm of a beat, by a melody; humans can understand as a unique whole a long sentence where the crucial word, which retrospectively supplies the key to the total sense, is placed at the end. A musical phrase depends on its final notes, and even a relatively long movement in a symphony can depend on its last harmonies. For time as for space—albeit to an unequal extent—the absolute domain is given with life, and the superstructures that the various techniques add to it are grounded in this primitive donation.

The significance of these considerations for the problem of finality can be easily understood. If we were to understand the structure of time according to the schema of classical physics, the idea of finality would become a pure absurdity, like the telepathy toward the past of Stapledon's Neptunian. Finality is incompatible with a series of actions ordered step by step. There is a "step by step" inherent in time, at least in the macroscopic time of physics in which one instant ceaselessly succeeds another. Although even a physical body can move here and there within a proper domain of space and return to its point of departure, where it eventually rediscovers and modifies its own traces, a physical body is swept away in time without return. Once space and time are fused, the irreversibility of time renders meaningless the return to the point of departure in space. The "universe line" obviously excludes every recurrence and every finalist organization. But, as we have already noted, the domain of absolute survey does not have the same structure as the space-time of physics. Here, in contrast, the absolute survey of space entails (with some supplementary difficulties) the absolute survey of time. Already, organically and psychically, I do not live exclusively in the present. I am always in the process of accomplishing an action or a labor that simultaneously anticipates the future and modifies the sense of the past. Despite the "step by step" of the succession of instants, which manifests the underlying reign of physical realities and aggregates, I do no enter the future with closed eyes. The present is not a block from which my free activity launches itself into the void. To some extent, we can choose our path in time, as we can decide between various itineraries in the surveyed space, thus avoiding future obstacles that can be identified with various symbolic procedures. To avoid taking a train on Saturday because, the next day, there would be no connection to the desired destination is not essentially different from the avoidance of an obstacle by an animal that alters its trajectory before it stumbles on this obstacle. The detour on the basis of innumerable virtual trajectories is equally possible in time.

Once completed, the chosen path has something definite about it. From that moment onward, the other possible paths I could have realized have a twice-imaginary pseudo-existence. In pure space, if I arrive at an impasse, I can return to my point of departure and restart; the method of trial and error is appropriate. In time, I can retrospectively alter nothing more than the sense *(meaning)* of my past conduct with the sense I superimpose on it. Nevertheless, the historian can hypothetically reflect on what could have been. And he has to do so, if it is true that every historian investigates what could have been to understand what was.[3] To understand the path that an animal takes—if one is not a dogmatic behaviorist who only seeks to use causes for explanation—is always to see it on the basis of "because otherwise." (The animal passed by here because otherwise it could not have attained the goal.) "Uchronia" is always more difficult to imagine than "utopia," and it is always more artificial; yet it is not impossible or absurd. Every comprehensive science is always based on the possible, on "utopia." Every social or individual history is always based on "uchronia." Utopia and uchronia are conceivable only through the notion of a domain of absolute survey.

d. *Choice and work.* Provided we set aside the problem of the "I," the substantialized unity of the domain, it is not difficult to show that the absolute domain is a necessary and virtually sufficient condition of freedom and work-activity. We have already seen that the two notions of freedom and work are indissociable.[4] Freedom and work presuppose, on one hand, the vision of a value or an ideal to be realized and, on the other, the choice of the means of realization. The two opposite notions (equally correlative to each other) are those of causal determinism a tergo and pure functioning, without ideal or possibility of choice. An automaton fitted with an artificial retina or cortex, like the photoelectric screen we have described, or even fitted with homeostats like those of W. R. Ashby, can neither aim beyond the actual nor choose its means. Innumerable automatic devices, which make use of "information" on perforated tapes and are already employed by life insurance companies, are essentially designed to select, to sort, to "choose," and they discharge their task better than a man could, who is replaced by them as much as possible and is guided in his conscious choice by the "automatic choice." But of course, the automaton "chooses" only according to the assembly realized in advance by the engineer. The most sophisticated device can only sort and differentiate; it cannot truly choose. An ordinary scale will "know" better than I can which of two objects is heavier, but I am the one who chooses the heaviest or the lightest according to the needs of the moment, which are a function of my reference through my field of consciousness to an order of ideal values. I am the one who chooses to use a scale as a pure means. The machines that verify the caliber of bearing balls by placing the "good ones" in one cabinet and the "bad ones" in another can only choose according to a predetermined assembly. The engineer incorporates the true choice into the assembly itself.

Only a domain of survey can choose, because the two objects to be differentiated exist together and distinctly in the subjective field, and because these objects are referred back, through the thematic and signifying character of true forms, not to an extremum, but to an optimum. An automaton does not have freedom of choice and, correlatively, it does not work (except in a entirely metaphorical sense).

Work proper always consists in establishing and improvising bonds and not in operating *according* to preestablished bonds. It consists in the "assembly" (in the active sense of the term) of bonds and not in operating *according* to an assembly (in the passive sense of the term). Cerebral work proper and cerebral fatigue probably imply that the neural cells are the initial supports of the assembly improvised by consciousness. The mental norm is transformed into a psychic "task"; this task tends in its turn to be transformed into material physiological bonds that function automatically. The act of choice becomes an organ of automatic sorting. Fatigue does not appear at the first, mental stage of the pure aim. It disappears at the third stage, when the mechanism is established. It is inherent to the second, psychobiological stage, because consciousness is then literally incarnated, serves as an improvised bond, and constitutes a unified system, perhaps removing energy from neural cells in accordance with a well-known principle: the interaction between a system's elements diminishes their individuality.

Because the absolute domain is the principle of every bond and not the outcome of bonds and of the assemblage of parts, it alone can work. Once built, a calculator supplies the "right" result much more reliably than humans; but the machine's cogs or circuits are simple proxies of improvised cerebral bonds. What we call "control" in machines like ENIAC or MARK, that is, the center that guides the opening or closure of circuits, is obviously nothing more than a control in the second degree, passive relative to the will of the manipulator.

A very interesting case is the one where the authentic choice made by absolute survey allows (through accumulation) the passage from disorder to order and, thereby, an inversion of entropy's normal direction of evolution toward a maximum. "Maxwell's demon," who choses the rapid molecules or the molecules of a mixture to make the heat pass from the cold body to the warmer body or to rediscover the component bodies by starting from a mixture, necessarily presupposes a domain of absolute survey. More generally, where we notice a rise in entropy (as in the biological order), we have to presuppose the existence of domains of absolute survey. This is a new and crucial sign in favor of the fact that the status of absolute domains underlies both organic and psychological phenomena. It is impossible to replace "Maxwell's demon" by a machine with automatic choices. Given that a machine can only "recognize" molecules through the physical or chemical action they perform on its organs, the molecules' interesting traits disappear in this very action. This restriction applies to organic machines qua machines. If we were to imagine Maxwell's demon, for example, with eyes like those of the multicellular organism, then they would only be able observe the molecules through their photoelectric effect, and the obtained information would become obsolete before it was useful. Only an absolute domain, in which knowledge is primary and independent of the observation that results from the interaction between individuals, in which knowledge is identical to autosubjectivity, in which the component parts of the system are not observed but seized in the absolute unity of the system, can resolve the problem of the rise in entropy.

Norbert Wiener noted that it is impossible to observe a given system (a star, for example) that would not obey the same thermodynamics we obey and whose entropy would move toward a minimum instead of going toward a maximum. In an interesting mental experiment, Wiener imagines a star that attracts light instead of radiating it. We obviously cannot observe this star, because we can observe the light that arrives but not the light that leaves. Continuing the utopian exercise, we can imagine an intelligent being B for whom time moves in the opposite direction to our time. For us A, every communication with B would be impossible or at least profoundly distorted: "Any signal he might send would reach us with a logical stream of consequents from his point of view, antecedents from ours. These antecedents would already be in our experience, and would have served to us as the natural explanation of his signal, without presupposing an intelligent being to have sent it. If he drew us a square, we should see the remains of his figure as its precursors, and it would seem to be the curious crystallization—always perfectly explainable—of these remains. Its sense would seem to be as fortuitous as the faces we read into mountains and cliffs."[5]

The following objection can be raised against Wiener: the living organisms, winding up entropy, realize in fact the conditions of his utopia, and yet we can observe them, and even observe them by means of a material system (photography, cinema) that obeys ordinary thermodynamics. But this objection is superficial and fails to capture the most crucial and interesting point. It is more accurate to follow N. Wiener's suggestion and to conclude that we cannot in fact observe them as living and acting qua organisms with finality. In the organism, finality is not, properly speaking, causality in reverse or the evolution toward minimum entropy; it is an "absolute survey of time" indifferent to its thermodynamic sense, because the end ideally precedes the means that precede it in the order of the actual. This is precisely why the scientific observation of organisms systematically misrecognizes their finality and has the illusion of causally explaining a finalist action that should be understood by means of survey.

Every individual action, which is independent of the normal entropic evo-

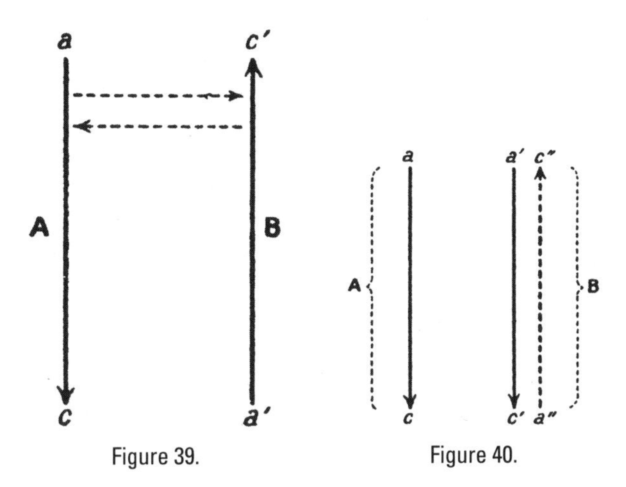

Figure 39.

Figure 40.

lution, is in the strict sense inobservable; and the "causalist" scientist who is confronted with a living organism is indeed like A, who explains the square that B traces by referring back to causes instead of understanding it as a sign, because of his failure to adopt the same direction of time as the being he observes. A living being understands *another living being only by placing itself in the finalist perspective and not in the perspective of mechanism or thermodynamics. Let A and B be the two beings whose thermodynamic time is opposed (Figure 39). What is antecedent for B appears consequent for A who observes B, and vice versa. B at* a' *decides to trace a square to emit a signal to A, and from* a' *to* c'*, the traces of the square are gradually erased by the rise in entropy. But because A goes from* a *to* c*, it sees the square emerge progressively and naturally without presenting any possible meaning other than that of a physical phenomenon determined by natural causes. Let us suppose now that A and B are cunning men, devious diplomats, conducting politics outside their rival countries. They obviously subsist within the same thermodynamic time (arrow in dashes), and now the antecedents* a *of A correspond to the antecedents* a' *of B. But B can have a secret project that will be revealed at the moment c', a project that B prepares with various maneuvers from the moment* a'*. These maneuvers are therefore consequent to the project's general theme. And everything takes place as if an arrow* a"c" *(Figure 40), which A cannot observe, combined—thanks to the action of conscious calculation—with the arrow* a'c'*, which A can observe. If events erupt, for example, at* a' *(events resulting, in reality, from B's* a"c" *project), A may very well attribute them to natural causes and fail to discern in them a* *sign of B's intentions, exactly as in Wiener's utopian example; he may treat the traces of the square drawn by B in the "future" as a natural phenomenon that has only causes and no sense. Consider a historical example: Churchill recounts in his* Memoirs *that he summoned American troops to Ireland from the beginning of 1942 to prepare the landing in North Africa. This arrival was therefore ideally "consequent" to the projected landing. But it was so indecipherable to the adversaries that it did not risk betraying the secret of the operation.*

So the absolute survey of consciousness and an inversion in the direction of entropy and in the course of thermodynamic time have similar effects. However, if A were very wily and shrewd, he could see through B's maneuvers and surmise his intentions from the first incidents, because, through his own consciousness, he would also be independent of the thermodynamic direction ac *and could enter into an ideal synchrony with* a"c" *while physiologically living in ordinary physical time. Every living being "knows" beyond what it "observes," though observation is always easier than knowledge and always risks obstructing the intuition that the nature of consciousness makes possible. Every living being is at once inside physical or thermodynamic time and outside time and the ordinary evolution of entropy.*

e. *Autoconduction and finality.* At the end of this analysis, we can conclude that the notion of absolute domain contains the key to finalist activity. We have been able to show that all the notions we can quickly extract from the constellation of finality are connected, directly or indirectly, to the very notion of such domains: work-activity, which aims for a definite optimum in the various orders of values; not simply spatial but temporal organization, which dominates the subordinate causal step-by-step chain and regulates it; coordinated possibilities, allowing choice and freedom; invention at all levels through the passage from forms to senses and from senses to forms, which not only permits the regulation of step-by-step chains but creates the means or the auxiliary constructions.

A single point remains to be clarified, and it is pivotal, not only for psychology, but also for metaphysics and even theology. The finality at issue here is a "harmony-finality" and not an "intention-finality." In other terms, a domain of survey is not a keyboard at the disposal of a distinct "subject" or "mind," the pianist, so to speak. The keyboard is capable on its own of autoconduction, and what appears in the universe of common sense as the intention, project, or goal of a man who speaks to us about what he desires for tomorrow and what he does today to prepare for it is the expression of a primary harmonization in this man's consciousness. Like perception's mise-en-scène, in which the subject seems focused on what he perceives, intention-finality, "spoken" finality, is a secondary technique that should not be transposed into the primary nature of finality.

Were it otherwise, it is obvious that we would merely have a pseudo-solution, which would not in fact avoid infinite regress. Even when it is an intelligent activity dominating the plane of perception, consciousness is not distinct from the intuitive or symbolic domain on which it is exerted. Cerebral equipotentiality gives the impression that the mind is detachable from the brain it uses (whether this brain is damaged or not). In reality, once the brain is totally destroyed, the "user" vanishes. It is often inevitable—we have done it ourselves—to personify the "surveying unity" of an absolute domain, to realize the division into unity, on one hand, and into multiplicity, on the other. But we should always remember that we are only dealing with metaphors here, because survey is "absolute," without "distance."

When, in the place of the pseudo-finalism and pseudo-spiritualism of Anaxagoras, whose Nous *is nothing more than a blind motor force, Plato wanted to define an authentic finalism in the* Timaeus, *he was inevitably forced to separate the Actor (the Demiurge) and his domain of activity, the World, and even to further separate the created World and the ideal Model. Erudite commentators puzzled over the exact meaning Plato attributed to this myth and whom Plato's true God was, the Demiurge or the Good. But Plato did nothing else than faithfully describe the very structure of every finalist activity. By definition, if we do not turn the mind into a simple fluid, we have to decompose it into acting subject and domain of action, and the latter into actual domain and domain of possible ideals. On the cosmic plane, the Platonic Demiurge is the* x *or the active unity of experience: "[There is] work," just as the Good is the value that defines the ideal pole of this same: "[There is] work." It is even more difficult to determine the extent to which Plato believed in his myth, that is, realized this double "division," because he probably did not know it himself. At any rate, the notion of absolute domain should allow us to conserve from this myth just enough not to relapse into Anaxagoras's pseudo-finalism, in one form or another; at the same time, we will not take seriously the division that transposes harmony-finality into intention-finality.*

It is the merit of Gestalttheorie *that it escapes every temptation to divide things, because, according to this theory, order and harmony are spontaneously constituted by a pure equilibrium, without a distinct active subject or a distinct order of value. Regrettably, however,* Gestalttheorie *is just a pseudo-finalism. This pure harmony-finality is no longer a finality at all. It is remarkable that Leibniz, by virtue of his dynamist principles, could have applied avant la lettre a kind of* Gestalttheorie *to metaphysics and theology: by a "divine mathematics in which the determination of the maximum takes place," the essences and the possibilities in the divine understanding pass into real existence according to their weights, which are conjugated with virtual existence. Leibniz believed he could speak of divine freedom and of God as final cause, the moral and not simply metaphysical perfection of the world. But truth be told, we do not see how this extremal dynamism is still a finalism. In reality, God plays no role in this matter; he is simply the site of possibilities, just as space is the site of existents. Leibniz's metaphysics is less mythical than Plato's, but it rests on an inadequate description of finalist activity.*

Leibniz's conception is not far from Hume's: "A mental world or a universe of ideas [such as the divine understanding, the site of possibilities] requires a cause as much as a material world or universe of objects."[6] *And if we reply that the different ideas that constitute the Supreme Being's reason fall into order on their own and by their very nature, "why is it not as good sense to say, that the parts of the material world fall into order, of themselves, and by their own nature?"*[7]

To take the conceptions of Plato, Leibniz, and Hume on cosmic finality for what they are (a simple magnifying glass that reveals the way in which finality in general can be conceived), we can say that the Platonic conception deploys the best phenomenology of finality. If Leibniz conflates dynamic equilibrium and finality and degrades the divine understanding into a pure ineffective site, Hume—to the extent that Philo is his authentic mouthpiece—is so preoccupied with avoiding infinite regress (which is laudable) that he pauses the analysis too early and does not see the essential difference between a collection of material objects juxtaposed in physical space (the parts of the material world or the objects scattered on the physical surface of a table) and a set of forms or of ideas in a domain of absolute survey (the universe of the ideas of divinity or *the view* of objects on the table). If my table is in disorder in an apartment I never enter, the objects that clutter it will have no chance of putting themselves in order. As soon as my gaze falls on this table, however, I can order the idea-objects that constitute the absolute surface of the seen-table according to the sense of my aesthetic, theoretical, social . . . activity. And yet there is here no risk of infinite regress. It would puerile to believe in a kind of "overconsciousness," "overperceiving" and "overwilling," which would once again see the seen-surface of the table and decide to tidy it up. But an absolute surface should at least be reached for there to be order and finality.

In this instance, the fundamental illusion lies in believing that the surface of the seen-table differs from the material table simply by a sort of illumination. According this metaphor, by turning on the light in the room—which I can eventually do without being present in it, through an external switch—I will not alter the order of the objects on the table. But consciousness, knowledge, and self-survey are not akin to an illumination; they are the presence of a primary mode of bonding, which subjectively exists as an absolute domain and is objectively manifested as equipotentiality. The seen-objects are no longer *alongside* one another like the material objects; they form part of a unitary system that acts unitarily. To believe that the seen-objects will continue to exist and to act as material objects, capable at the very most of blindingly pushing one another, is to arbitrarily dissociate consciousness's mode of existence and its mode of acting, even though the mode of existence is nothing other than an abstraction of the mode of acting. The subject, the surveying unity (the Platonic Demiurge), is the unitary action in the present participle; it is the Acting [*Agissant*], the present participle substantialized. The material objects as clusters or machines operate only according to their structure and their step-by-step bonds. From *fonctionnement* [functioning], a substantialized present participle, which would be *Fonctionnant,* does not spontaneously emerge; clusters or machines act as "subjects" only in sentences. By contrast, an active pole emerges from the unitary action inherent in the absolute domain, a pole that appears to be opposed to the passive domain that undergoes the tidying-up according to a sense.

12

The Region of the Transspatial and the Transindividual

It is impossible to understand the world of space, time, and individuals, unless we consider it a kind of limit to a natural world or region, but of an entirely different nature than our visible world, a region in which the spatial or temporal "step by step" does not reign and in which the analogical "same" and the numerical "same" blend together. This is a region of essences, ideas-forms, and mnemic themes. The main trait of the domains of absolute survey is to make these ideas-forms, these transspatial essences, visible in the observable geometric structure of the spatiotemporal world. A sensory cerebral area (or rather, its real and autosubjective counterpart) can be compared to a one-way mirror, which, on one hand, receives the physical images of the observed objects and, on the other, reflects the essences of the transspatial world that correspond to these objects.

The region of essences and themes should not be situated in some mythical geography, like the one that amused Plato. It can be reached through positive descriptions of a certain number of psychological facts, all of which reveal the same structure.

a. *Mnemic evocation and invention.* Bats had been using ultrasound to explore obstacles for millions of years when P. Langevin designed his receptor-and-emitter device to explore marine obstacles. To produce ultrasound, the bat does not use piezoelectricity like the industrial device, but it has to resolve the same technical difficulties (e.g., emitting wave trains short enough that they would not scramble the echo).

We do not generally observe the same degree of resemblance between two independent inventions as between two evocations of the same memory. If I have for a second time a complex idea that I'd already had, it will likely resemble its first image in my mind more than its image in the mind of another man. Once discovered, invented, and become a memory for me, an idea will be more easily at my disposal and will also have a better-defined nature than a universal essence. Ordinary reminiscence is easier than Reminiscence in the Platonic sense. But even this is not always true, as experimental psychology has demonstrated, and in every case, there must logically be a reason for the resemblance. The resemblance of two geographically independent inventions, like the resemblance of two evocations of a same memory, has to find its reason in a transspatial "nature."

It is very typical that the "actualist" and "existentialist" conceptions deny at once (1) constituted and unconscious individual memory; (2) specific memory, the reason for the resemblance of one human to another or one animal to another of the same species; and (3) the world of essences and values independent of our whims. In their own way, these conceptions acknowledge the solidarity of the three orders of reality. That this negation is insupportable is, in any case, clear for individual memory and specific memory. We have to account for the fact that two swallows or two humans resemble each other. The reason does not reside in a kind of material "negative" from which they are drawn, like two mass-produced objects; but this does not exempt us from searching for and finding the reason elsewhere. The same logical obligation equally holds for the resemblance of two inventions in all domains, and more clearly in domains in which the discoveries obey such rigorous norms that they can be rigorously identical (e.g., mathematics or technology).

We should not succumb to the illusion of believing that the resemblance of two inventions can be explained in the same way as the resemblance of two phenomena of macroscopic physics. There is also something typical about the form and evolution of a delta, the form and evolution of the meanders of a stream, of a cumulus, of a volcanic eruption, because common nouns are used to designate them. Yet this "typicality" is secondary and derived; it is sufficiently explained by the play of causes at work in these phenomena, which always operate in the same way. Strictly speaking, an invention has no cause; it differs by definition from a functioning. The resemblance of two inventions cannot be placed without contradiction on the same plane as the resemblance of two erosions. Even in the domain of classical physics, by ascending from cause to cause, we arrive at the "nature" of the primary physical beings at issue, which, strictly speaking, do not have a cause and respond to a "type." The resemblance of two iron molecules cannot be explained in the same way as the resemblance of two cumuli; a molecule conforms to a norm that can be analyzed mathematically, but not—strictly speaking—causally. The application of mathematics differs profoundly depending on whether micro- or macroscopic physics is at issue. In macroscopic physics, mathematics allows us to follow the deduction of causality from primary types to derived "types." For instance, the geographer–mathematician will calculate the formation time of a particular type of delta from the flow of sediments, the nature of the shoreline, and the adverse action of the waves. In microscopic physics, by contrast, the mathematician does not have to deduce the effect-phenomena from the cause-phenomena; an iron molecule is obviously not formed in the manner of a delta.

So long as the choice was only between the irrational negation of the absolute empiricists or existentialists and the mythical substantialization of essences or types in a transcendent site, it was possible to hesitate—although, in any case, a mythical image of reality is always worth more than a logical absurdity. No doubt the "swallow" type or the "human" type, the essences or values in their atemporal status, cannot be imagined as an Idea enthroned in the heavens and contemplated admiringly by beings that strive to imitate it. But the attentive observation of the facts of memory and of the true modes of reminiscence can give a positive meaning and value to the old Platonic reconciliation of invention and memory.

b. *Mnemic subsistence.* Experimental psychology has demonstrated that in the vast majority of cases, humans only recall meaning. When it bears on meaningless syllables or figures,[1] the effort of memorization essentially consists in relying on auxiliary meanings or on various "expressivities" (rhythms, spatial clusters), expressivity being here as everywhere else a "nonexplicit meaning." When memorization progresses, auxiliary meanings and mnemotechnical "tricks" are often abandoned. But the active memorization of a meaningless material cannot do without the detour through a "signification" or an "expressivity." And the same holds for the "memory" that passes through notional "detours."

For instance, a subject who has to memorize the pair of syllables viz-hus *relies on the analogy with* vicious, *then abandons the auxiliary term when the pair becomes familiar to him.*[2] *It is very doubtful that a mechanical or photographic memory, an absolutely pure memory (pure of every meaning or expressivity), could exist.*

Naturally, experimental psychology cannot resolve the problem of the mode of mnemic subsistence on its own. But even if memorization and recollection rely on "meanings," one cannot easily claim to relate the mnemic subsistence (between memorization and recollection) to a phenomenon of mechanical inertia, especially when so many other facts (especially those analyzed by the Wurtzburg school, Freud, Burloud, Ellenberger, in dreams and spontaneous associations) uncover the peculiar life of mnemic spheres almost directly and when the analogy between invention and memory corroborates the presence of the transspatial. This subsistence has to be of the same order as the subsistence of non-mnemic meanings, that is, of essences.

In their phenomenological status, essences are eternal; they are endowed with ubiquity (an inventor can invent and perfect his invention in any part of the world). In the region of essences, the similars are identified, while in the region of existents, similar beings can be numerically different. The status of mnemic, individual, or specific subsistents is quite similar. Psychological, individual memory possesses a kind of ubiquity: I can evoke my recollection as well in Asia or America as in Europe; a kind of eternity: a recollection is detached from time, it can return to me at any moment, it is "eternal" up to my death; last, it escapes number in the sense that the same memory can be actualized any number of times, mutating slightly at each moment. Organic memory is even closer to essences. Its ubiquity is less relative than that of individual memory: multiple embryos of the same species can develop simultaneously, very far from one another. Its eternity is also less relative: a specific memory can endure for millions of years—altering somewhat at each ontogenesis—so long as a single couple of the species survives. Despite everything, however, a difference always subsists between specific memory and the authentic eternity of the essence: "red" could reappear even if, at a given moment, no "red" existed in the universe and if "the species of reds," so to speak, were temporarily extinguished; the specific memory of a vanished species is annihilated forever. Where is the specific memory of dinosaurs? Finally, specific memory eludes number in the sense that it can pertain to any number of individuals without being universal and indivisible like the essence.

We can thus glimpse the profound reason for the effort toward sense that every memorization makes. The actual can only escape time by participating as much as possible in the status of the essence, by profiting in some sense from its eternity to gain a precarious and limited immortality. An idea, a memory, is a hybrid between the eternal and the actual. If I have an idea at some point, this idea that is mine and ephemeral like me will also be universal like the essence it aims for and, in principle, eternal in the same way. A memory is always necessarily an idea. Bergson's assertion that every actual automatically becomes a memory is false. The actual becomes a memory only when it is imbued with "sense," which renders it incorruptible. Our recollections subsist only when they are transmitted in time through the eternity of essences.

c. *The action of resemblance.* For a long time, philosophers emphasized that the impossibility of understanding the action of resemblance within the strict horizon of the pure actual. How can a form or a being A evoke the form or being B that resembles it, because resemblance is a relation of psychological or mental order that presupposes two presented terms, and because, by hypothesis, A alone is presented when it evokes B? Everything becomes clear when we see the action of resemblance as the inverse of the successive actualizations of the same idea or the same memory. In this case, the same idea (numerically and analogically) engenders a multiplicity of similar actualizations. In the case of the action of a resemblance, an actual form, seen as a corresponding idea or essence, evokes the other possible actualizations of the same essence. It should be stressed that from the form seen by the retina to the form seen as an idea, there is no "passage," despite the upward arrow of the schema (Figure 41), because to explain the passage, an action of resemblance has to be introduced, and this would obviously displace the problem without resolving it. Every perception is thematic and directly grasps the *eidos* in the structure-form. Within a schema, the "detour" through the transspatial and the transnumerical inevitably looks like a progression; but the essence-in-itself and the essence incarnated in a form are in fact identical by virtue of the transnumerical nature of the essence. In the same way that mnemic subsistence is a participation in the eternity of the essence, the action of resemblance is a participation of the actual in the law of numerical identity of similar beings, which reigns in the transspatial region.

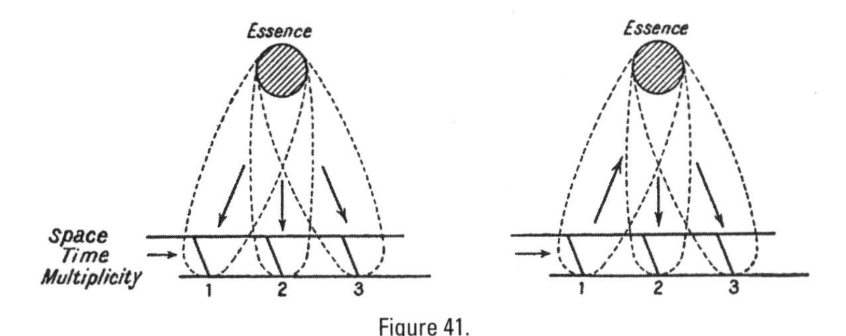

Figure 41.

d. *Imitation.* Imitation also poses an insoluble problem for "actualists" and mechanists. How can the fact that A hears B sing explain the fact that A hums in his turn the heard tune? How can the sight of his mother's smiling face provoke the smile of the child? The auditory or visual cortical zone is not traced on the motor zone. The key to the solution resides in the remarkable isomorphism between imitation and memory. (1) We only imitate what we understand; we only imitate meaning, just as we only memorize meaning. (2) Imitation amounts to surmounting the spatial diversity and distance as well as the difference of individuality between the imitated and the imitator, just as memory amounts to surmounting the temporal diversity and distance between a primary experience and its recollection. (3) Imitation, like memory, reveals a difference in height between the threshold of recollection and the threshold of recognition. Experiments, especially on animals (Köhler) and on children (André Rey), have shown that we only imitate what we nearly managed to invent; but a threshold difference—like the difference between the threshold of recognition and the threshold of recollection—allows us to identify the right solution and to imitate it, just before managing to invent or "name" it. The imitation–memory analogy thus incidentally reinforces the invention–memory analogy. (4) Memorization is a function not only of the signifying but also of the important, the valuable in general; the well-known effects of mnemic reinforcement of affectivity and moderate emotion, which are themselves due to the intimate union between affectivity and the sense of values, have their counterpart in the fact (stressed by Dupréel) that we particularly imitate what we admire. Hence the enormous psychological and social role of models. The admiring emotion is to imitation what emotion in general is to memory. (5) We can add that imitation in familial or social education prolongs the action of heredity and organic memory. The child imitates his parents; his psyche is modeled on their psyche, just as the organism of the child was formed by the same specific potential that already formed his parents. (6) The almost indiscernible use of the two procedures (mnemic and imitative) in biological reproduction is a very striking confirmation of the analogy between memory and imitation.

We have already underscored the fact that the reproduction of a multicellular organism always appeals to a memory, and to a memory that is not the property of a tissue or the operation of an organic structure, because epigenesis has been confirmed experimentally and because this organic memory creates the tissue and structure. But there is another form of reproduction, which we can designate as "reproduction by self-replication": the reproduction of viruses, bacteriophages, chromosomes, and genes. In this type of reproduction, there is a splitting of an actual structure and not an epigenetic reconstitution of a structure. Is such self-replication a mechanical tracing [calquage]*? This hypothesis is as implausible as that of the mechanical nature of psychological imitation. According to recent observations (especially those of Pease and Baker in 1949, with the electron microscope), a gene is an already-complicated structure in which a long chain of proteins resembles the central column of a spiral staircase, whose steps would be represented by nucleic acid molecules associated with the specific protein. We realize that this staircase can be easily cut in two, resulting in two similar parts. But the gene still has to reconstitute its primitive length; and we do not understand how architectural elements can be multiplied by a mechanical tracing or by purely physical phenomena of "resonance" when this length is reconstituted. The reproduction of the chromosome presents other difficulties. Because genes are schematically stacked in the direction of their length to form the chromosome, the longitudinal splitting of the chromosome during mitosis can be understood as cutting all the genes in two, in a single stroke. But matters do not seem so simple. The "position effects" of genes (Sturtevant), of "translocation" (Bridge, Müller), seem to show that the chromosome is not only the sum of genes but also acts as a whole. Reproduction by mechanical tracing of such a complex whole is altogether implausible; and Goldschmidt, who criticized the autonomist conception of genes and defended the conception of the chromosome as a genetic unit, is forced to postulate "the ability of the chromosome to reproduce its own image by division or by recreation of its* *likeness."[3] The truth of the second term of the alternative seems all the more probable, because the asexual reproduction of protozoans or the reproduction of cells, which is also a splitting of* actual *structures, has absolutely nothing to do with a pure and simple division. There is something epigenetic about it, and it forms a bridge between the mode of reproduction of viruses and that of multicellular beings. For instance, a shelled protozoan is forced to protrude out of its shell a cytoplasmic hernia in which skeletal elements migrate, all formed in the protoplasm of the mother cell. The protozoans' divisions take place according to very varied modes. They seem to be essentially mnemic phenomena, just as much as the reproduction of multicellular beings. So in the case of genes or viruses, there is a reason to hesitate over the mnemic or imitative nature of the division. This hesitation is instructive; it demonstrates that "imitation" cannot be a mechanical tracing but implies the action of a typical resemblance and presupposes a detour through the region of the transspatial as much as epigenetic reproduction presupposes a detour through hereditary mnemic potential.*

Haldane and other authors have attempted, without insisting too much on this suggestion, to link the problem of genes' and viruses' reproduction by self-replication to the fact that a particle in quantum theory has no definite individuality. It would be impossible to say which of the two copies of a gene that is being reproduced is the model and which is the copy. Given the dimensions of a gene or a virus, this parallel is just an analogy. But it is valid because the lack of individuality, which is clear on the plane of space and time, is in both cases a sign of the two "mitigated" individualities' intimate connection with a transspatial "type."

All these facts thus exhibit the same schema. They have this in common: they set in play a resemblance within the actual, without invoking a mechanical tracing to explain it. Memory without engrams, the action of resemblance, imitation without tracing—all of this is contrary to the laws of ordinary physics and cannot be explained by them. To account for these phenomena, we have to resort to transspatial themes or essences. The resemblance of two actualizations of a single memory requires the idea of a mnemic theme; the organic resemblance of two individuals of the same species requires the idea of a specific potential. By the same token, the resemblance of organs between two very distant species, the resemblance of these organs to our tools, indicates that all these similar actualizations are "financed" by something else, which is situated in the transspatial region. In the psychobiological order and in the order of invention in general, the *number* of actualizations seems indifferent; it seems to cost nature nothing and is only limited by the occasion or the available material of realization. Viruses, protean-viruses, unicellulars, and the embryos of superior organisms, like the mnemic evocations or the evocations of ideas, can be divided and multiplied to infinity. The indifference *of the number* of actualizations perfectly responds to the indifference *to the number* of essences and themes. The number of existing circles is arbitrary relative to the "Circle," just as the number of swallows is arbitrary relative to the "swallow" type: essences and themes are both transindividual and transspatial. It does not take much (a contact with an inductive substance, a hair that the experimenter tightens more or less around the egg) to decide that there will be two individuals instead of one. A minimal actual coincidence is enough to evoke afresh a haunting memory. When a scientific or philosophical idea is in the air, it does not take a great event for it to hatch in one mind after another, each believing itself the first and the only to discover it. As often happens in the history of science, the fact of biological "determination," which at first appeared to be such a curious and aberrant phenomenon, proves to be extremely general. It resembles the call of memory, the action of the circumstances or of chance on invention. All these phenomena are inconceivable without the duality of the spatiotemporal world and a region of essences and memories. "Causes," circumstances as insignificant as the various agents of biological, mnemic, or inventive "determination," cannot account on their own for the immense developments they initiate.

The active existence, the development, of individuals is a continuous "suction" they perform on the transspatial world, a "nutrition" in the most general sense of the term. A being feeds on "sense" more profoundly than it feeds physiologically or materially. Subjectivity, which is the reality of every being, is merely a series of acts of comprehension, of sense. Biological, psychological, and mental development is nothing other than a continuous annexation of mnemic properties and riches. We cannot really describe human beings without describing their ideas, their memories, their assimilated experiences, their vocation, and their aspirations—in short, the whole "transspatial" that enriches them. It is impossible to turn biological development and, a fortiori, psychological and mental development into a simple functioning in space. A development has no causes that can be localized in space-time. The domain of space-time is nothing more than a limit; it cannot even really contain existents, because their subjectivity "straddles" the two regions and because their instantaneous structure and even the whole series, moment by moment, of their instantaneous structures is just an abstraction.

More generally, the whole constellation of phenomena we had to describe in our overview of finalist action requires an invisible region to double the visible world. In more exact terms, because we seem to subordinate the invisible world to the visible world, "the visible world is always the invisible world realized here and now."[4]

It is through language above all that humans dwell and move within the transspatial and the transtemporal. We mean humans in general and not only a few speculative thinkers. Businessmen, politicians, managers, and utilitarians, as much as metaphysicians and mystics, are elsewhere than where they are. It is impossible to head toward a goal, to utter meaningful speech, to listen to or look at another being in space, without leaving space. When two men chat on the telephone, the exchange and the mutual understanding cannot be reduced to material transformations like those in play in the telephonic emitter and receptor. The current of information that circulates between the two men leaves space-time at each of its extremities to reach the domain of essences and significations, where barriers between individuals disappear.

The human head is reversible: it can alternatively speak and listen; but it is not reversible in the same way as an electric machine. Its reversibility is more subtle, because it is in communication in both directions, back and forth, with the transspatial. Humans live much more in the world of symbols than in the material world, and if efficacy and utility were criteria of reality, the world of symbols would be more real than the material world. Travelers on a train are not bodies resting on a bench; they are beings haunted by invisible goals, who cannot stop speaking of them to one another or to themselves. Here "world of symbols" does not mean "world constituted by symbols" but "world explored by symbols." Human significations are not superimposed on a senseless world. Signification is nothing more than a technique of sense, that is, of the invisible part of beings. Words do not create senses or essences; they are possible only because there are biological and mental types and species.

13

The Levels of the Transspatial and Finalist Activity

We still have to examine, if possible, the internal architecture of the region of the transspatial. But this crucial clarification should be made first: by situating essences and mnemic themes beyond the space-time of classical and relativist physics, we are not denying them all the traits of "form"; quite the contrary. The Greek term *eidos* has very fortunately the double sense of "idea" and "form." If form (idea-form or theme-form) is not in space and time, it is nonetheless spatializing and temporalizing. The idea-form has a relation to space-time, because it accounts for geometric forms in space and in time. The mnemic themes of a living species control its anatomic forms and the temporal melody of its development. The transspatial form is "beyond" space-time, but precisely because it creates it. It has been clear for a long time that a space or a time that would be consistent with its definition as *partes extra partes* would not in fact be a space or a time but a sort of pure multiplicity that could not constitute a universe.

We have to elude the "step by step," to survey it, to grasp it as a mode that can be defined abstractly. An absolute domain is constitutive of space-time because it differs from a physical survey by its double relation with the region of the transspatial, on one hand, and with the "I" or *x* of individuality, on the other.

The region of the transspatial is not opposed to the space-time of classical physics in an abrupt way; it presents several kinds of subregions that become less similar to space-time and its content as they "distance themselves," as it were, further and further from it. We can distinguish at least four regions:

1. the region of actual consciousnesses with the extensive sensations and the *specious present,*[1] limited ubiquity and eternity
2. the region of individual psychological memories

3. the region of specific organic memories
4. the region of essences and values

To these we can add the "region" of the "Sense of senses," the transcendent Unity, the supreme Logos, the unnameable Tao that no longer has a nature.

By passing from one region to the other, we move further and further away from the laws that reign over the physical world: step-by-step causality and the numerical diversity of similar beings. The "mnemic" regions 2 and 3, considered together, represent a sort of contamination of the transspatial by the individuals of space-time; they contain essences that are "appropriated," "specified," and "converted" into themes or types. Organic memory resembles, on one hand, individual memory and, on the other, essences and values.

The problem of psychophysiological "parallelism"—it would be better to say, if usage had not consecrated the first expression, the psychophysiological "correspondence"—now appears as a particular case of a much more general problem. At level 1, the parallelism or the correspondence is still very clear: actual consciousness escapes space-time only insofar as it is not subject to a *simple location* (according to Whitehead's expression) and insofar as it cannot be matched to a punctual element. Nevertheless, it is very much subject to the unfolding of space-time. I have a sensation or an emotion at this moment and in this place, and an alarm siren stirs at the same time the residents of a city. I dominate a nonpunctual yet very limited spatial domain; I dominate isolated moments, but I am swept away by time. If I pronounce a sentence, the intuition of its sense, like the "I," temporally dominates the physical unfolding of this sentence; but afterward I can fall asleep or become distracted. When the sentence is somewhat long, the psychological sequences of subordinated senses closely parallel the spatiotemporal and physiological events that accompany its emission. However, if I firmly maintain the general sense, and if I am not suffering from syntactic or semantic aphasia, then this sense will escape the parallelism. In a complex behavior that requires multiple steps, the parallelism in the details of execution between physical or physiological phenomena and psychological experiences is quite close (e.g., the memories of a statesman are replete with details of time and place as well as daily psychological impressions). But the general sense of this behavior is only vaguely and very roughly connected to the historicogeographical frame. It dominates this frame from on high. The psychological life of a conventional Epicurean, dragging himself from meal to meal, closely parallels his physiological life. His emotions correspond perfectly to the gastric secretion of the appetite, then to digestion. But the psychological life of a statesman, leading a great nation in a great war, does not really parallel what physiological observation registers minute by minute in his organism. A psychological "pointillism" would provide an entirely false idea of this life.

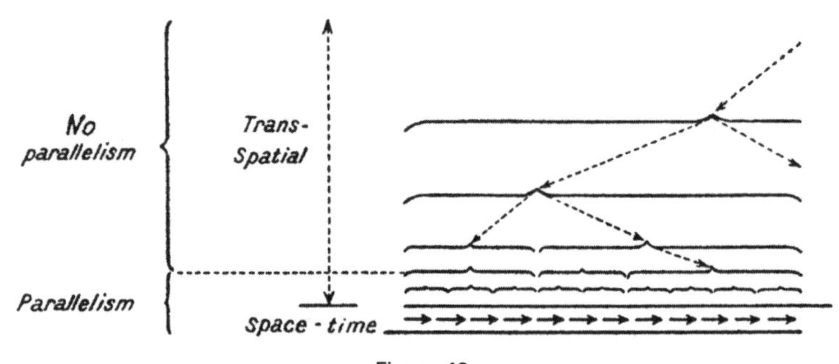

Figure 42.

In the senseful and finalist activity of a superior organism, beneath the levels where the great themes of actions detach from space-time and step-by-step causality, we find a level of realization of details on which the minor themes that regulate these details correspond very closely to the spatiotemporal phenomena observed by science. A complex behavior can always be schematized with a hierarchy of curly brackets, which represent themes whose final lower stage, while remaining within the transspatial (because it is a question of themes), corresponds very closely to the spatiotemporal unfolding and is modeled intimately on the small accidents of causality that the influence of higher themes "regulates." The small technical accidents of an action are normally attenuated by the play of numerous, increasingly higher, encompassing stages detached from space-time. It is thus quite clear that, depending on the stage we consider, we are at times struck by the parallelism, at others by the absence of parallelism. The absence of parallelism and, therefore, the reign of the transspatial are nevertheless fundamental and essential, and the structure of finalist action (with absolute survey of themes) can still be found at the final level of realization. The reign of spatiotemporal step-by-step causality (the domain of horizontal arrows) is nothing more than a limit to the domain of vertical curly brackets. Whether the bracket is immense or miniscule, it represents the same fundamental mode of absolute survey.

In an obscure novel, Abel Hermant (probably influenced by the peripheral theory of emotions) had the bizarre idea of deploying a physiological vocabulary to describe, instant by instant, the emotional and sentimental reactions of his characters, or at least to describe them without referring to their sense. The novel is barely readable (except where the author strays from his method).[2]

It is true that if we undertook the opposite challenge in a novelistic or historical tale, that is, if we only described the most general "surveying" of ideas or themes without ever descending to the "nearby" stages of spatiotemporal unfolding, we would produce nothing interesting for readers who love concrete descriptions. Neither novel nor history would in fact exist any longer, nor even a tale of any kind. The non-negligible interest of a phenomenology or a theoretical treatise would at least remain, valid for all times and all places, or at least valid for a large zone of social and technical culture.

Reality lies in fact between these two poles, thanks to layered themes or shock absorbers that adapt the transspatial region to the region of space-time. On one hand, they allow the incarnation of essences and values by a descending action and, on the other, let the accidents of space and time modify in their turn, by an ascending action, the themes that survey them.

It is typical that psychophysiologists grappling with patients and not with theories, like A. Meyer, Goldstein, or Kantor, were unable to use pure behaviorism in practice, nor pure mentalism, nor the abrupt body–mind dualism, that they had to rely on more "unifying" conceptions, which they sometimes had difficulty defining for lack of a clear conception of the relations between the transspatial and the spatial. Kantor notes, "Psychological events may be regarded as the larger field situations of which biological activities, howsoever essential, constitute only components."[3] Kantor pushes things to the point of paradox when he denies the brain any role in the perception of objects, renewing the Bergsonian paradox that places the brain in the middle of a world of images and not images in the brain. But what is false for the perception of real objects is quite true for the apprehension of ideal beings or of very general themes of action.

It would be absurd to reduce the guiding idea of a great enterprise like Operation Neptune to a few states of consciousness, which parallel the temporary physiological states of its architects. An "ideal" is a value assumed by a living being, just as an "idea" is an actualized essence. As a result of their appropriation by an individual, the ideal and the idea have certain localizable and even measurable psychophysiological effects. But it would be quite strange to reduce the ideal and the idea to these few effects. Countless men can sacrifice themselves for the same idea. Will it be claimed that they sacrifice their lives to some physiological processes that take place in one corner of their own brains? Will it even be claimed that they sacrifice their lives to some well-localized and dated psychophysiological processes? And by what miracle could these localizable processes be designated as *one* idea or *one* ideal common to all? Consciousness is truly consciousness only insofar as it contains more than its instantaneous perceptible content: it exists above all in the invisible world, which it explores.

The levels of the transspatial are multiple. The passage from a visible and observable world, outlined by the emissions of photons and by elementary energetic interactions, to the invisible world of informing themes or ideas is continuous. The parallelism is neither absolutely true nor absolutely false. It becomes increasingly false only as we "ascend" further into the invisible world, distancing ourselves from the instantaneous structuration of the "observables."

Like parallelism, and for the same reasons, the Orphic–Platonic conception of two worlds is at once true and false. It is true only in the "highest" regions of the inobservable world. But there is no abrupt, Orphic or Gnostic opposition between two worlds. Man is not double, even though his being is situated at all levels and even though he sometimes almost touches, despite his materiality and his animality, the unique Logos and the unnameable Tao. From the most elementary chemical phenomena of his organism to his highest ideal, there are so many intermediaries that they cannot be separated. The "soul" has not fallen from the heavens into a body; it cannot leave the body through ecstasy or asceticism to voyage and return to its homeland. But an idea or an ideal, by transforming the body into a simple subordinated instrument, realizes in fact the Platonic ascesis without resorting to myth.

The myth par excellence always consists in taking the term "world" seriously in the expression "the invisible world," in transforming the region of the inobservables into a sort of *Umwelt* imagined on the template of the biopsychological *Umwelt* and in which the soul can journey and contemplate. It is well known that Christianity, for instance, oscillated more than once between the two conceptions of man, the Orphic–Platonic conception and the much more unifying Aristotelian conception. This oscillation is understandable, because the two conceptions are simultaneously true. Man is one in the sense that, on the last level, the visible body and the primary consciousness are a single and same reality: the parallelism is perfect for the good reason that the body *is* the last level of observed organic subjectivities. And yet, when man wishes, he can also become a demigod who makes use of his body as of a disdained instrument. There is no separation between two worlds, no obstacle, no astral spheres, no empty immensities, no oceans of darkness guarded by evil angels as in Gnostic myths. The transspatial domain's many levels are not separations but degrees. Human consciousness can elude its individual, biological, and even psychic limits. For animal consciousness, instinct is a sort of "mandatory mission." But human consciousness can choose the idea that will sweep it along. An idea is not the subject of contemplation. The choice of an idea is the choice of a mission, and the level of our work is the very level that our soul attains in the transspatial.

14

The Beings of the Physical World and the Fibrous Structure of the Universe

The main obstacle to the adoption of a finalist or neofinalist philosophy, to which so many facts converge today, stems from a deep-rooted prejudice, according to which the visible and tangible matter is *all the same* more real than senses, ideas, and values. The end of mechanism with quantum physics and wave mechanics has not been sanctioned by a corresponding change in our vision of the world. The very expression "wave mechanics" attests to the persistence of a mechanistic and materialist vision. For after all, why should we continue to speak of "mechanics," that is, of "machines," apropos the schemas of the atom as they have been figured since L. de Broglie, Schrödinger, and Dirac? The expression "wave domains" or even "wave organizations" would be more justified.

Perhaps this prejudice has its roots in images inherited from the scholastic age. We are thinking of the mineral, plant, and animal "Kingdoms," the mineral Kingdom being the support of all the others. A philosopher as modern as N. Hartmann continues to take these superposed "Kingdoms" seriously, and by dividing each of the two terms of the Cartesian dualism, he has systematized this theory of Kingdoms. He has distinguished four foundation-levels (physical, biological, psychic, and mental), each superimposed on another *(Überlagerung),* either through *Überformung* [superformation] (an organism is composed of physical elements) or through *Überbauung* [superposition] (psychic life is constructed on organic life). N. Hartman defines what he calls the "categorial laws of dependence."[1] This categorial dependence is valid only from the lower to the higher layers. The lower categories are thus "stronger." For example, causality, which can be physical as much as biological and psychic, is "stronger" than finality, which does not reign in the physical world. The categories of lower layers are the "existential foundations" *(Seinsfundament)* of the higher ones but are "indifferent" to them. They make the superposition possible but do not favor it. The higher layers cannot subsist without the lower ones, but the latter can subsist without the former. While breaking with materialist monism, like N. Hartman, who also believes that each level has a categorical *novum,* the philosophers of emergence (Lloyd Morgan, Alexander) have similarly retained from materialism the notion that the world is a kind of multileveled structure whose ground floor (matter, *Grund,* space-time) alone is solid. Alexander (who unlike N. Hartman was writing under the influence of the theory of relativity and before wave microphysics) went so far as to turn space-time into the only genuine God, because deity emerges from it as a final category after value, consciousness, and quality.

Contemporary science invites us to adopt an entirely different view of things. The visible and tangible, spatiotemporal and "material" world is no longer a point of departure for science, a fundamental given; it is, instead, a point of arrival and something whose construction can be followed from what is not visible or tangible, from what is not spatiotemporal or material. The molecules and atoms of nineteenth-century physics were the "bricks" from which the world was built. It would be quite superficial to believe that the contribution of contemporary physics has consisted in going further in the search for these constitutive "bricks." The protons, neutrons, electrons, photons, mesons . . . have not simply replaced the atoms and molecules; they are not the new bricks of reality. These elementary particles are not particles that would exist as such in the finished edifice. Instead, they are like cells or organs in an organism or words in a sentence. Living cells can be cultivated in vitro, words can be considered in themselves and defined in a dictionary or employed in a phrase; but the organism or the phrase is not a simple structure whose entire reality pertains to the elements. Quite the contrary, if we wanted to push the analysis further, for example, by decomposing a word into letters and each letter itself into small segments, all reality would vanish. By the same token, even if it enjoys a certain individuality when it produces a photoelectric effect, the photon or the meson does not preexist as a distinct particle in the atom that emits it by passing from one energy level to another; it is integrated into a unitary domain. The general traits of life or of language pertain to cells and to viruses as much as to metazoans, to short sentences as much as to long ones. The general traits of absolute domains pertain to the beings of microphysics as much as to the beings of psychology.

Since Stanley's discovery, on one hand, and since quantum physics, on the other, it has become impossible to represent the universe—the real universe of individual beings—as made up of a series of superposed layers, the lowest bearing the others. The universe has, instead, a fibrous structure in time, and each fiber represents the continuous line of an individualized existence.

The virtual immortality of protozoa requires the life of an actual protozoan to be represented by a long "fiber" climbing back to the very origins of life. The divisions of reproduction and the unions create bifurcations or interweavings of "fibers" but do not hinder their continuity. Because it seems very likely (after the discovery of ultraviruses) that unicellular beings derive from large organic molecules, the "fiber" can climb back much higher, up to the very origin of the real universe. This schema is obviously limited to the protozoa. The somas of multicellular beings are mortal, but they derive from immortal germinal cells. By definition, none of the cells alive today has ever perished. Each dates back to the very origins of the universe. But unlike the germinal cells, each somatic cell has to confront the impasse of death. On the other hand, the schema of "fibers" also applies (albeit less easily) to actual physicochemical individuals. They too date back to the origins of the universe. They have no doubt undergone countless avatars: they lost and regained electrons or photons, but these avatars are of the same order as the exchange of nuclei in the course of the combination of protozoa, for example. The schema does not become impossible to apply until we reach the most elementary "particles" of microphysics; these particles have an indeterminate individuality, and it is impossible to "track" their identity in a domain of interaction, as though far from being the fundamental bricks of the construction, these "particles" were less "substantial" than complex individualities.

The lower "faunas," that is, the chemical species, do not constitute the fundamental layer. The higher organisms are indeed "made up" of cells, molecules, and atoms (by *Überformung*), but not in the same way that a house is made up of bricks. Instead, the cells or molecules are "possessed" from within by an individuality that managed to colonize and organize, according to a thematic unity, a collection of other individualities often produced from its own division. This "possession" should be conceived on the model of the possession and the reciprocal capture of psychomnemic spheres and not as the relation of a brick to a wall. Physical beings are in no way more real than the higher organisms; they cannot serve to explain them or to make them intelligible.

Instead, the opposite is true. For, if microphysical individualities are absolute domains just like psychobiological individuals, then the description of the domains we experience directly (a visual sensation, for example) can help us understand them.

The risk of error is great, but not as great as the risk of the apparently similar approach of physics over the last three centuries: it consisted in drawing conclusions about physical elements from tangible bodies and from artisanal and industrial machines. For example, Newton writes, "We know by experience that some bodies are hard. Moreover, because the hardness of the whole arises from the hardness of its parts, we justly infer from this not only the hardness of the undivided particles of bodies that are accessible to our senses, but also of all other bodies. . . . The extension, hardness, impenetrability, mobility, and force of inertia of the whole arise from the extension, hardness, impenetrability, mobility, and force of inertia of each of the parts. . . . And this is the foundation of all natural philosophy."[2] Newton's example is hardly encouraging. The falsity of his inference is clear. And yet it seems much more adventurous to infer the character of physical beings from a visual sensation or a human activity than to infer the hardness of atoms from the hardness of stones. Contemporary physicists who, taking microphysical indeterminism seriously, spoke of the "freedom" of the electron and connected it to human freedom have not had very good press. Nevertheless, the audacity of this new parallel is in fact not as great as the audacity of Newton and mechanistic physicists. These latter believed they were simply inferring from the whole to the part that is homogenous to this whole, but they went illegitimately from "molar" and statistical properties to individual properties. This movement is equivalent to conflating in biology the physical and geological properties of sedimentary limestone levels with the properties of the individual mollusks that constituted them. By contrast, contemporary physicists who strive, like Bohr, Jordan, de Broglie, and Eddington, to connect microphysics and biology or psychology, the indeterminism in the atom and human freedom, remain at least within the order of individuality (despite the obvious audacity of this reconciliation). They respect the sense of the "fibrous structure" of the universe. The limestone-shelled animals that constituted miles of sediments may not look a great deal like human beings, but they resemble them (because they are living individuals) more than they resemble a sedimentary layer. We should not therefore be intimidated by the irony with which the "freedom of the atom" is greeted.

Nothing is easier than ridiculing this thesis of "freedom" by transferring to the atom all the accessory effects of the freedom of superior organisms and saying, for example, that "if humans are free to marry or to remain celibate, it is because some key-electrons in their brains can make or refrain from a quantum leap." But if the term *freedom* is troublesome, it is enough to replace it with the term *activity,* which is its exact synonym. To speak of the freedom of the atom or the atomic element amounts to saying that the atom is an "agent" and not a "functioning." Expressed in this way, the thesis loses its scandalous character, because it is precisely the quantum of *action* that is the origin of infra-atomic activity's indeterministic character. Likewise, it would be absurd to misuse Jennings's and Mast's observations on protozoa to the point of attributing the calculations and emotions of a human hunter to an amoeba. It is nevertheless true that the most meticulous observations have highlighted the general traits of psychic behavior among protozoa, against Loeb's mechanistic theories: spontaneity, variety of means, persistence of the act up to the desired end, reaction of the organism as a whole.[3] The essence of freedom, even in human beings, does not consist in producing movements without reason or "doing what one wishes." It consists in having at one's disposal a domain in which an infinity of virtual possibilities become simultaneously visible, in which space-time is not a network of points-instants tied together step by step but an idea-form in which genuine actions and not pure functionings take place, actions that take place according to a norm and use the virtual possibilities as means. In short, as we have already stressed, freedom is inseparable from finalist work-activity. It is effectively possible to rediscover these traits, or a good portion of them, at the level of microphysical domains. It is the whole "constellation" of finalist activity and not only freedom that we can discover in them.

1. The paradoxes of microphysics stem from the insufficiency of the ordinary notion of space-time, according to which the successive instants simply mark the progress of a functioning and are not tied to the dynamic character of an absolutely unitary action.[4] Quantum indeterminism derives from the existence of a quantum action.

To consider a very clear comparison made by Whittaker, let us suppose that a pure note is produced by an organ pipe. Its frequency v *is very low such that the number of oscillations per second is small. The key that controls the emission is supposed to act very rapidly. If we ask at which instant the note of frequency* v *was sounded, we cannot give "a precise answer, since the sound would actually extend over the interval of time during which the key was down. In order to obtain as nearly as possible a note sounded at a precise instant, [it is necessary to] shorten its duration as much as possible . . . but, by doing so, we cut short the train of oscillations . . . [and] the sound will no longer produce on the ear the sensation of a definite pitch: it can no longer be described as of frequency* v*."*[5]

Action (energy multiplied by time) is homologous to work-activity in that, within the space-time of conscious work,[6] there is no possible *simple location* of the constitutive movements that are thematically subordinated to the unity of action.

2. To represent the behavior of a photon or an electron, microphysics associates with the corpuscle a continuous field that represents its various possibilities of manifestation. It should not be assumed—this error is commonplace—that the associated probability wave is meaningful only for particles in large numbers.

If, for example, we produce light interferences with an intense light source emitting a number of photons and a screen riddled with holes, the zone of interference can be calculated by means of associated waves. Up to that point, we encounter nothing extraordinary. But if we decrease the intensity of the source until a single photon is emitted at a time, interference fringes nevertheless appear on the photographic plate, *where the photons arrive one after the other, producing localized photoelectric effects.*[7]

Everything happens as if *one* photon were able to explore the whole screen and its multiple holes and not only to follow a linear trajectory. Physicists have resolved this paradox by considering the wave associated with a single photon as the probability wave for its location. The unique photon does not then have to pass through one of the holes to the exclusion of others, and we do not need to ask through which hole it effectively passes. The photon has no determined position at the interior of the wave. "There is in some sense a 'potential presence' of the corpuscle at every point of the region of space occupied by the wave."[8] In fact, it appears as a corpuscle only at the moment of photoelectric interaction. It is difficult to avoid the impression that at least an analogy can be made here to the internal ubiquity of domains of survey.

3. It is ostensibly more difficult to find in microphysics the equivalent of the "end" that characterizes free activity, which always seeks to attain a final optimal state according to a norm. But reduced to its essence, this character simply amounts to the following: whereas there is a deterministic functioning within a system when the modifications of this system are proportional to the motor energy applied to it from the outside, there is "activity" when the modifications cannot be linked to causes a tergo but are defined by a final state in the most general sense of the term *final*. This second case is that of intra-atomic changes in their specificity. No doubt there is also a causality through accidental impetus in the "life" of the atom. If an atom is bombarded with accelerated particles or if an incident photon dislodges an electron or makes it leap to a more external layer, what happens to the atom is due to a causality a tergo or at least to an accidental and external reaction. But, on the other hand, we know that the structure of the atom in no way resembles the structure of a planetary system, in which the trajectories of the planets are established at various distances from the attractive center, distances that result purely and simply from the equilibrium of given masses and speeds. The quantum of action structures the atom in a well-defined way and endows it with a certain "type," to which it returns or tends to return *despite* the external incidents and unpredictably (except when atoms are considered in large numbers). The superior organisms are also subject to some accidental causality a tergo, although they are essentially capable of a proper, "regulative" activity that is consistent with an ideal norm.

S. Stebbing[9] mocks the reverend J. H. Morrison, who is obviously somewhat pressured by the need for pulpit eloquence to establish that "at the heart of reality, there is a divine activity, an urge, a desire for self completion," and who uses as an argument the fact that physics rejected dead matter and "introduced in its stead 'action' as the ultimate physical reality."[10] *S. Stebbing responds, "That* 'action'*—in the physicists' sense, i.e.,* momentum times distance*—could be regarded as equivalent to, or in any way analogous to, a desire, 'an urge towards the attainment of the ideal,' 'a will to live,' is simply absurd."*[11] *The irony that greets the "ideal" of the atom is exactly of the same kind as the irony that greets the "freedom" of the atom. It is not grounded in more rigorous reasons. If we take "ideal" like "freedom" with its highest human traits, then it is obviously absurd to speak of an "ideal" of the atom. It is as absurd as attributing to the amoeba the emotions of a big-game hunter. But if we consider the essence of finalist action in opposition to the essence of functioning, we will discover a more profound resemblance between the self-regulation of an atom—whose "excited" electrons return to their primitive orbit by reemitting a photon at an unpredictable moment—and the self-regulation of an injured organism than between an atom and a planetary system, in which the distances and speeds are regulated by step-by-step influences.*

In a living organism, "formation" (in the active sense) is indissociable from form. A living being is never "fully assembled"; it can never confine itself to functioning, it incessantly "forms itself." This is precisely why, for living beings, the problems of origin and formation are indissociable from the problems of nature. Contemporary physics forces us to say the same thing about physical individualities. An atom is not a fully assembled mechanism that functions. It is incessant activity; it continually "forms itself." An activity or an active formation is indissociable from a norm. If the "type" of a definite atom cannot be understood as the simple presence (persisting through inertia) of a ready-made structure, then it can only be a normative type. A hydrogen atom ceaselessly "makes itself." It cannot "be there" once and for all, any more than a living being or a social institution. Because it is nevertheless possible to designate it as a hydrogen atom, it has to obey a norm, and its nature has to be a *physis* in the etymological sense of the Greek term.

4. Individual existence, as it appears in a finalist manifestation, is indissociable from activity itself; it is not the existence of a substance that can be inactive.

The notion of "functioning" implies that there exists in the first place a static, material, or substantial structure that moves but that can also remain at rest. In contrast, true action, free action, implies that no material or mental substance is posited at the origin, because acts would either be inherent to it as properties and therefore would not be acts or would be pure "emergences," which do not need to be tied back to the substance and as a result would not be *its* acts. Leibniz's metaphysics allows for no more freedom than Democritus's philosophy.

It is remarkable that physical beings fulfill in this sense the most profound conditions of both freedom and existence. They are not static structures. The atom in contemporary physics does not have an "existence at instant *t*"." It can only be defined as a certain type of action in which time is integrated in the indivisibility of action. G. Bachelard vehemently emphasized this for a long time: "The problem of the structure of matter must not be separated from the problem of its temporal behavior. . . . Wurtz justifies atomism by invoking the ancient argument that it is impossible to 'imagine movement without *something* that moves.' To this argument microphysics might well retort that it is impossible to imagine a thing without positing *some action* of that thing."[12]

Collingwood, on the other hand, notes, "An atom of hydrogen possesses the qualities of hydrogen not merely because it consists of a certain number of electrons, nor even merely because those atoms are arranged in a certain way, but because they move in a certain rhythmical way."[13] For the atom as for the living being and the conscious being, we "cannot separate what matter is from what it does."[14] "The old idea was that first of all a given piece of matter is what it is, and then, because it enjoys that permanent and unchanging nature, it acts on various occasions in various ways. It is because a body, in itself or inherently, possesses a certain mass, that it exerts a certain force in impact or in attracting others. But now the energies belonging to material bodies not only explain their actions upon each other, they explain the extension and the mass of each body by itself. . . . So far from its being true that matter does what it does because first of all, independently of what it does, it is what it is, we are now taught that matter is what it is because it does what it does: or, to be more precise, its being what it is is the same as its doing what it does."[15]

This is certainly a paradox, because to act, one must be, according to the structure of Indo-European languages and to the structure of common reason. But it is a paradox that runs exactly parallel to the one we find in the phenomenology of work and human freedom,[16] where we are forced to investigate the curious formula "work and you will exist" or Lequier's formula "make, and by making, make yourself."

We cannot therefore say that matter is *mens instantanea,* in accordance with the Leibnizian formula. A physical element is nothing if it is instantaneous, if it is not a certain prolonged rhythm of activities. So long as we believe in the traditional material "substance," time can be conceived as an empty dimension along which the substance is passively borne. But when the traditional concept of matter is replaced with the concept of activity, time can no longer appear as an empty and foreign frame; the time of action is inherent to this action as a temporal melody. This amounts to saying that it can only be conceived as the mnemic rhythm of activity. The physical rhythms are indiscernible from a memory.

The main difference between physical beings and the most complex organisms does not probably derive from the instantaneity or the absence of memory in the former but from a lack of detachment of this memory, which in physical beings is always inherent to the rhythm of activity, which is only ever "the form in time" and does not constitute a transspatial "reserve" clearly detached from the actual. In human beings, memory constitutes "other Is" that enrich the actual "I." In all organisms proper, organic memory constitutes specific potentials that can be reincarnated in innumerable individuals.[17] In physical beings, no enrichment of this kind is found. The semisubstantialization of activities into "mnemic beings" does not take place for physical beings. So, contrary to the prejudice of materialism and of the "philosophy of layers," it can be said without paradox that the material world is less substantial, more "pure spirit," more "Ariel," than the organic and psychic world. This crucial fact aside, a perfect isomorphism exists between the finalist activity of higher organisms and the activity of physical beings. To speak of the "freedom of the atom" is not a laughable blunder of philosophers who are poorly informed about science and followed by preachers in search of apologetic arguments. On the contrary, we have to expand the thesis and speak not only of the freedom but also of the finalist and regulative activity of physical individualities.

Despite the general isomorphism between all activity-forms, and although there exists no physical matter, fundamental reality, solid ground, substance, *materia prima,* relative to which all of the other realities would be ephemeral superstructures (what sweets are to sugar), it is clear that the activities-forms beneath which no other activities-forms exist must have a very special status. Contemporary science's rejection of *materia prima* does not entail the suppression of the problem of *forma prima* or *activitas prima,* because the problem lies in the epithet *prima* and not in the noun that precedes it. Organisms present themselves as hierarchical, colonial Empires. And so "noncolonizing colonies" exist at the final level of these Empires. The cells in an organism do not resemble bricks in a wall, but they are indeed subindividualities. As a result, when we reach the final level, we come up against a paradox. On one hand, the facts prove that the general properties of absolute domains are conserved; on the other, it is impossible—unless an infinite regress is accepted in this instance as well—not to arrive at a domain that is no longer colonial, that no longer has dominated subindividualities. This seems to contradict the very notion of domain, where *dominus* must have "inferiors."

The problem appears under a practical form: the difficult interpretation of "conservative" principles. In modern physics, since the theory of relativity, the old principles of the conversation of energy and the conservation of matter have given way to a more general "conservative" principle in which mass and energy are one. An activity conserves itself; a substance does not. What does the conservation of an activity mean? In the past it was thought that the conservation of energy could be clearly conceived as a substance-traversing-time. This clarity is illusory. Today we should try to conceive the conservation of an activity without the false clarity of the idea of a persisting material substance.

At our human scale, "activities" are generally not conserved. While the conservation of matter is a quasi-intuitive given, or can be easily inferred with simple arguments (like those of Lucretius), the conservation of activity contradicts our intuition, our habits, which are formed through the experience of macroscopic activities. The activity of higher organisms incessantly passes through highs and lows, through alternations of *fortissimo* and *pianissimo,* like a Beethoven symphony. We go from work to rest, from wakefulness to sleep; we can mobilize our energy or remain idle.

This property of higher organisms is bound up with their colonial and composite nature. Systemic unity, as we have already seen, is directly correlated to the interaction between the system's component elements. The more intense the interaction, the more the individuality of the components is effaced for the benefit of the whole. Thus, in a composite system, a transfer of activity from elements to the system is possible, and vice versa, a transfer that corresponds to the increase or decrease in interaction and in binding forces. If we only consider the macroscopic activity of the whole, then we will have the impression of an intermittent activity, even though there is merely a change in the balance: individuality of the system ↔ individuality of elements. Because activity is synonymous with freedom, we can say that in a system that loses its unity, the elements reassume their own activity and their freedom, which had been partially mobilized when the system acted as an individual. The sum of activities or of energy can thus remain constant in the universe, despite the intermittences of higher activities. The physical systems in which the interactions and the internal binding forces are extremely energetic even seem to "produce" the particles they emit or release, because these particles have no distinct existence at the interior of the system where they interact intensely.

The form of organisms' activity always derives, in the last resort, from the placing of the individual *x* in circuit with the transspatial: essences, values, organic or individual memories. But the energy of this activity, that is, its quantitative and measurable aspect, cannot derive directly from this placement-in-circuit. Countless experiences prove that the "form" and "energy" of an activity are largely independent. The same light energy contributes to the growth of plants with very heterogeneous forms; the same energy from the same nutrients nourishes the most varied animal activities. Hence the materialist thesis according to which the elements of things are more fundamental than the complex structures that exploit these elements. But we can understand the conservation of the universe without reverting to the materialist thesis. A special status has to be attributed to "final domains," which are colonized by the others and do not colonize. But this special status is poles apart from what classical materialism imagined. The "final domains" are the least substantial of all domains, they are pure activities; paraphrasing the expression Descartes applies less fittingly to the soul, we can say that they "always act." They are uninterrupted activity; they cannot rest or sleep like higher organisms. They cannot even temporarily demobilize their elements, for they have no elements to demobilize. They are a pure unity of action without a subordinated multiplicity. They have neither a structure nor even, strictly speaking, a form. They only have an activity-form; and the spatiotemporal domain and the metaphysical "transversal" can no longer be dissociated in them (even ideally) as in the other domains. The two are now one. They lack a detachable memory, and they have no need for one, because they never have to take up again the thread of their uninterrupted activity.

It is worth emphasizing that, interpreted in this way, the principle of conservation has absolutely nothing to do with a rational principle. The necessity of attributing a special status to "final domains" is relative to the verifiable existence of complex domains with variable activity; it is relative to the verifiable existence of a certain conservation of energy and action in the universe. This is not at all a rational necessity. It is inconceivable that "final domains" would cease to act and would continue to exist; but it is perfectly conceivable that they would cease *at once* to act and to exist or begin at once to act and to exist. In the present phase of the known universe, a conservative principle holds roughly true: the elementary organisms are incomparably more stable than the complex organisms. But nothing prevents us from conceiving of a phase in which the total quantity of energy would vary and in which elementary beings would appear and disappear. If we want to understand the current situation of the universe, we have to admit that elements are "always in circuit" with a transspatial. But it is not essential for the elements to be always in activity and to exist continuously in time. G. Lemaître's bold hypothesis[18] about the prodigiously "energetic" primitive atom that engenders, through radioactive fragmentation, the entire universe perhaps lacks boldness after all, because it continues to maintain the conservation of mass-energy. To believe that the scientific ideal is the mathematization of the universe as a conservative system is probably a prejudice. Its "fibrous structure" is the expression of lines of activity and not lines of subsistence. The subsistence of things derives from their activity; it is not required a priori by reason or virtue of a principle such as "nothing is lost, nothing is created." Activity in its unfolding is not subject to deterministic causality; development-activity depends on a deterministic causality in the world only for its initiation here and now. Why should it be subjected to this causality in its absolute appearance or disappearance? This idea is altogether meaningless, because a deterministic causality implies the interference of a multiplicity of elements. Furthermore, we can perfectly envision a phase of the universe or, according to Whitehead's expression, a *cosmic epoch*[19] in which we would witness the irruption of new elementary domains with creation of energy, just as in the current universe we witness the emergence of new complex organisms with approximate conservation of total energy.

Why even speak of an "other" phase of the universe? The universe we observe is expanding. The models of the universe in vogue today have an increasing radius, a decreasing density, and a constant mass. But several physicists (Hoyle, Lyttleton, P. Jordan) reject the postulate of constant mass and begin to envision models of the universe with variable mass and increasing radius, the expansion perhaps balancing a permanent creation of matter (one nucleon per cubic meter and per billion years). It is, moreover, possible to combine a denser original state (which several astronomical accounts confirm)[20] and an increasing total mass.

15

The Neomaterialist Theories

In a certain superficial sense, we can say that contemporary science has realized the hopes of the old materialism concerning the problem of life. We can say that the problem of the historical origin of life no longer arises. The appearance of life from a geologically "dead" world can no longer be considered as one of the forever-insoluble "enigmas of the Universe." The modes of emergence of complex organisms are far from being known, but the emergence of life, considered as an absolutely novel mode of being, is no longer a philosophical problem. There is no longer any reason to believe that from a chemical molecule to a bacillus, the abyss is greater than from a bacillus to a vertebrate. Physicochemical sciences and the sciences of the organism are much closer to each other than they were in the eighteenth and nineteenth centuries. They have already blended together in practice. The study of crystallizable viruses, genetic mutations, and large organic molecules in general attracts both chemists and biologists. One physicist after another, from N. Bohr to P. Jordan, L. de Broglie to E. Schrödinger, has his say on the problem of life.

This triumph of "materialism" is illusory. To affirm that microorganisms are molecules is to admit, in the same stroke, that molecules are microorganisms. The universe's "fibrous structure," made up of individual lines of continuity, is the capital fact underscored by all the recent discoveries. The physics of "individuals" enters into continuity with the biology of individuals. The living organism can no longer be reduced to a complex of physicochemical phenomena in the ordinary sense of the term, that is, into aggregate and statistical phenomena. Physicochemical phenomena certainly unfold in the organism and are used by it; but they are not the organism itself. One might as well claim to explain the chemical properties of the water or salt molecule through the laws of hydrography or oceanography. Today, the mechanistic[1] (or physicochemist in the classical sense) theories of life are nothing more than an archaism.

We will not dwell, therefore, on the old materialism or the physicochemist doctrine. In reality, their many contemporary representatives tend more and more to appeal to "neomaterialist" accounts, as we will define the term further on.[2]

For example, M. Prenant mixes with the customary accounts of mechanistic materialism, on one hand, "dialectical" accounts and, on the other, arguments drawn from the physics or chemistry of individuals. While claiming that zoology has become a "comparative biochemistry," J. Needham acknowledges that biology cannot be mechanistic in the strict sense of the term because electrodynamics and the atomic physics of quantum theories do not derive from the principles of classical mechanics. We find similar proclamations among other supporters of the physicochemical reduction: M. Werwon, Schafer, F. H. Marshall, E. B. Wilson, and so on.

Yet a neomaterialism that no longer appeals to ordinary mechanics or to the statistical laws of physics and openly admits the new fact of the physics of the individual, in continuity with the biology of the individual, is apparently possible. The expression "in continuity with" that we have used is readily neutral. But two extremisms (in the etymological sense of the term) are possible, because the line of continuity from the physical molecules to the higher organism has two extremities. One can argue with Whitehead, A. Meyer, and, to a certain extent, J. S. Haldane that the notion of the organism should be stressed and that the philosophy of the "organism" should dominate "physicist" philosophy. One can, in contrast, stress the notion of "physical element" and consider even the complex organism as secondary relative to the physical element or the individual with which it is in continuity. The obsolete ideal of reduction to the old physics or mechanics has left a subtle influence in the scientific atmosphere, and despite the radical mutation introduced by the physics of the individual, one continues to believe in a poorly defined primacy of the molecular and the elementary.

It is seemingly enough to explain the situation clearly to bring the irrational nature of this residual belief to light. Along the lines of continuity, along the indivisible fibers, there is development and not composition. Composition is at least always subordinated to development, as in the passage from the egg to the adult multicellular organism. The idea of reduction and of analysis was meaningful so long as one believed in the primary character of the phenomena of classical physics. This will no longer be the case when we realize that every individual organism is, as such, as primary (i.e., unanalyzable into aggregate phenomena) as any other individual. It is absurd to say that a protozoan is "in reality" a virus that complicates itself. It is equally absurd to say that a virus is "in reality" nothing more than a molecule. What is this "reality" in the expression "in reality"? In the wake of the modern theory of homopolar bonds, we can no longer say that the water molecule is "in reality" two hydrogen atoms and one oxygen atom, because it includes a zone of "absolute survey." How could we then say that the organism of a vertebrate is "in reality" an enormous molecule? If we are shocked by a generalized "theory of the organism" in the sense in which Whitehead, for example, understands it, it should be clear that the choice is between this theory and the theory of the "generalized molecule," as E. Schrödinger expounded it. Schrödinger's thesis contains the following truth: like the opposite theory, it insists on individual continuity. But we fail to see how it is favored by the rationalist ideal. The majority of biologists and physicists who have maintained theses similar to Schrödinger's realize the reversible character of their "physicism," and several among them would no doubt protest against the epithet "neomaterialist." The extremes often touch, and a "neomaterialist" is at times indistinguishable from a neofinalist. The epithet is nevertheless justified to the extent that these authors retain some measure of the ideal of reduction.

For several decades, the failure of ordinary physicochemistry led to the idea of interpreting the organism as a supermolecule, a supermatter, or a paramatter,[3] which obeys different laws than ordinary matter.

Benjamin Moore[4] considers the colloid as a kind of supermolecule produced by molecular affinities; and these differ from the atomic affinities that form the ordinary molecules but are of the same order. Its self-regulative properties and its faculty of reproduction depend not only on its structure but also on a special biotic energy. Rignano also appealed to a specific vital energy, although it is related to physical energies and is completely different from the "vital force" of classical vitalism.

Struck by Pictet's experiments in which animals refrigerated to −120 degrees came back to life, Max Loewenthal[5] considered life as the outcome of the persistent structure of a supermolecule: each cell is built by a complex and plastic brain that is in reality a unique and gigantic molecule. Such a molecule cannot vibrate and heat up like an ordinary molecule (Schrödinger elaborated this idea); it absorbs kinetic energy and conserves it in a latent state, in the form of intra-atomic energy. It may even be, as E. Montgomery suggested, that all the cells of the nervous system are in protoplasmic continuity and form nothing more than a single molecule.

A. Gaskell,[6] *on the other hand, suggested that the protons and electrons that can form the ninety-two kinds of ordinary atoms can also be united in combinations of a different and unknown type (he dubbed it "z-systems") to form the living matter by association with ordinary systems of particles. Strictly speaking, "z-systems" are not material; their only body is the ordinary atomic systems with which they are associated and to which they communicate the typical properties of life.*

But the progress of atomic and quantum physics gave a decisive impetus to these neomaterialist speculations, which, up until 1930, had remained quite arbitrary. Niels Bohr[7] was one of the first to perceive these new possibilities. He suggested that the quantum uncertainties can exist at the point of insertion of vital phenomena, which are refractory to statistical physics, and that to the principle of complementarity in quantum physics corresponded something equivalent in the biological order, because we cannot observe a living organism and experiment on it without killing it.[8] The label "neomaterialist" would poorly suit Bohr, who insisted on the specificity of life and did not share the reduction bias: "The existence of life must be considered as an elementary fact that cannot be explained, but must be taken as a starting point in biology, in a similar way as the quantum of action, which appears as an irrational element from the point of view of classical mechanical physics, taken together with the existence of elementary particles, forms the foundations of atomic physics."[9]

R. S. Lille, a biologist, invokes the atom's internal and individual activity.[10] This activity, which is independent of external influences (as the impossibility of controlling radioactivity and quantum "leaps" demonstrates), represents an enormous quantity of energy; and instead of being drowned in the statistical effects of ordinary chemistry, it can be manifested in the organisms—in their genetic mutations or even in their individual behavior and their unitary control. The most typical properties of organisms (progressive differentiation, subtle and often asymmetrical structure, spontaneity and selectivity) would result from intra-atomic factors, which become effective in the direction of the entire system and distinguish it from statistical systems. Vital activity would be none other than the direction and control of quantum actions or interactions from one atom to the other. Like N. Bohr, R. S. Lillie is close to both neofinalism and neomaterialism, despite the accent he places on the atom and atomic activity, because he admits that the constitution and "internal" activity of the atom to which the categories of space and time do not apply very well can be of the same order as the immediately intuited psychic activity.

S. C. Smuts,[11] W. Stern (for whom the atom is a "person" of a lower order), Ch. Eug. Guye, Lecomte de Nouy, Louis de Broglie, Bouchet (who insisted vigorously on the importance of the advent of a science of the individual), and A. Jakubisiak and A. Moyse[12] cannot be easily classified as neomaterialists. By contrast, G. Matisse, whose thesis was discussed earlier, clearly sides with the reductive ideal: "The organism is a kind of stereo-chemical super-molecule."[13]

P. Jordan has insisted above all on the probable kinship between the discontinuity of atomic reactions and that of genetic mutations, which can be provoked by shortwave radiations that perhaps only act at first on a single atom.[14]

Because E. Schrödinger attempted to inject more precision into this hypothesis, we will examine his exposition in detail. The second founder of wave mechanics (after Louis de Broglie) starts from commonplace considerations on the statistical nature of ordinary physicochemical laws, which as a result cannot apply to the most specific vital phenomena. "The arrangements of the atoms in the most vital parts of an organism . . . differ in a fundamental way from all those arrangements of atoms which physicists and chemists have hitherto made the object of their experimental and theoretical research."[15] The chromosome "may be suitably called *an aperiodic crystal.* In physics we have dealt hitherto only with *periodic crystals.*"[16] An organism integrates an enormous quantity of atoms in its physiological operation, but it is controlled by groups of atoms so small that they escape the laws of large numbers. These groups of atoms are the chromosomes, which contain the temporal as well as spatial total pattern of the adult organism in a sort of *code-script.*[17] This code-script would allow an all-pervading mind like the one Laplace envisioned to read in advance the entire future development, all of whose details are inscribed in the code in a one-to-one correspondence. It is executive and instrumental at the same time as legislative. It explains both the stability of the organism and the very rare mutations that trigger its evolution. The mutations are discontinuous like a quantum leap. The comparison between a mutation and a quantum leap is more than a comparison. Schrödinger's essential thesis is that the mutation is just a modification of the quantum state in the gene-molecule. To prove this, we can rely on a vital clue: the laws that regulate the rate of mutations induced by X-rays, as N. W. Timofeeff specified them, are remarkably simple. (1) The growth factor of mutations is exactly proportional to the dose of rays (which proves that the mutation is not a cumulative effect but an individual event). (2) If we vary the wavelength of the rays, the coefficient remains constant, provided the same dose in *r*-units is given (number of ions produced per unit of volume in a standard substance).

Apart from rare mutations, a gene's stability despite the thermic molecular agitation can be explained only because the gene is a molecule. A molecular configuration is stable by virtue of the same quantum principles that produce the mutations, but which normally interpose an energy barrier between one configuration and another.[18] Because genes govern the structure of the organism, the entire organism is then an aperiodic, stable crystal like a molecule at absolute zero; it escapes Carnot's principle and the progress toward maximal entropy and disorganization. It does not pass "from order to disorder" nor "from disorder to order." In other words, its order is not static like the order of the secondary laws of physics. It passes "from order to order," the fundamental order being that of the chromosomes, which not only conserves the organized structure but also establishes itself and grows by "extracting order" from the external environment, by "feed[ing] upon negative entropy."[19] The organism is therefore a "pure mechanism," like a planetary system without tides or a clock that functions without any friction or overheating and without any statistical degradation.

E. Schrödinger's conception is typically neomaterialist.[20] If we disregard the true schema it contains (the open recognition of the "fibrous structure" of the universe), then it can be easily critiqued, precisely in all of its materialist aspects.

a. It rests on the postulate that chromosomes and genes represent a sort of code-script. This postulate is more than contestable; it has already been proven false by a whole host of experiments. The scientific value of genetics is not in question, but after the experiments of E. Wolff, Baltzer, and his school, even the most dogmatic geneticist would not affirm that a "one-to-one" structural correspondence exists between the genes and the adult organism.

b. Let us even admit this postulate against experience. In no way does it explain the organic order in its specificity. It would explain the stereotyped conservation of a given structure but not the supple and constantly inventive regulation of the living being. The organism does not confine itself to enduring or retaining its order; it ceaselessly remakes it by perfecting it. Schrödinger is obviously disappointed when the logic of his system leads him to compare the organism to a pure mechanism, to a *clock-work*: "We seem to arrive at the ridiculous conclusion that the clue to the understanding of life is that it is based on a pure mechanism, a 'clock-work' in the sense of Planck's paper."[21] By this he no doubt means, in a negative sense, that the organic order *is not* a purely statistical order. The founder of wave mechanics knows better than anyone that a molecule, a crystal, or an atom is not a *clock-work* in the ordinary, kinematic sense of the term but rather a dynamic system; and Planck's paper to which Schrödinger alludes effectively opposes "dynamische und statistische Gesetzmässigkeit."

On the other hand, by asserting that the organism "feeds upon negative entropy," he implicitly ascribes to it a conquering activity and not a pure maintenance of order. But this amounts to saying that the interest of the comparison between the organism and a molecule or an a-periodic crystal is in the direction organism → molecule rather than in the direction molecule → organism. Whenever we find that it is plausible, says Schrödinger, that a "current of order" issuing from the molecule-chromosome may produce other ordered events, "we, no doubt, draw on experience concerning social organization and other events which involve the activity of organisms. And so, it might seem that something like a vicious circle is implied."[22]

Schrödinger's endeavor attests to the error of neomaterialism and the necessity of openly accepting neofinalism. The truly primitive order can only be founded on an essentially normative activity, of the same type as the psychobiological activity; we tried to define it in describing the domains of absolute survey, in which a metaphysical transversal dominates the form by endowing it with a sense. The purely material order and the brute persistence of a material order (in the substantialist sense of the term) are secondary phenomena relative to the primary order. By replacing the atom of matter with the indivisibility of action, physics has in fact shown the superficial character of an order that is nothing more than a brute persistence, and it is quite curious that Schrödinger failed to do in biology what he managed to accomplish in physics. Had he not argued his thesis with perfect clarity, we would have readily suspected that he misspoke or that we misunderstood him. For him, the molecular, "crystalline" order of the gene and the organism is an absolutely passive order, because it is in principle the x-rays, the cosmic rays, or the rare fluctuations of thermal energy that explain the mutations. If these mutations are favorable, they will be conserved in an equally passive manner by natural selection. Like a statue, the organism is modeled by a bombardment of physical particles. Neomaterialism thus intersects with neo-Darwinism, which we will examine further on and which also considers the organism as a passive structure.

For Schrödinger, quantum indeterminism plays an essential role in accidental mutations; it intensifies their accidental character. But, contrary to the opinion of Bohr, Lillie, and L. de Broglie, it plays no role in freedom and in activity at the level of consciousness.[23] Moreover, Schrödinger sees no other means of reconciling the intimate feeling of freedom with the determinism that reigns in the natural laws obeyed by our organism than the recourse to vedantic speculations on the identity of Atman and Brahman, "I" and "God." "I" am the one who controls the movement of atoms according to the laws of nature. There is only a single "I." The plurality of "Is" is an illusion. I (as Atman) obey determinism, but because I (as Brahman) created determinism, I feel free. This bizarre intervention of Hindu philosophy in Schrödinger's work is an artificial addition. But, in reality, it is logical vis-à-vis the error at issue. If the organism is just a passive order, then the unique and direct source of order can only be God himself, as in all deterministic systems. Starting from the consideration of individualities that are dependent on statistical laws, Schrödinger loses the genuine individuality along the way; he fails to take seriously quantum indeterminism in its positive aspect as an authentic activity. His case confirms what we have already observed: the psychological and "organic" interpretation—in the broad sense—of the individualities of contemporary physics is not a fantasy of the incompetent metaphysician but a capital truth that cannot be dissociated from everything else.

It is too indulgent for neomaterialism to claim that, after all, it posits the same thesis as neofinalism, albeit in the opposite sense. When one compares the organism to a molecule, one quickly succumbs to the old errors of mechanistic determinism. When one continues to conceive of the physical individual without activity, without freedom, without subjectivity, and without normativity, the kernel of neomaterialism's truth (the recognition of lines of individual continuity) is itself quickly lost and misunderstood. The opposite risk of naive anthropomorphism is less grave and can be avoided. It is enough not to invert the neomaterialist thesis purely and simply, not to define the atom, the molecule, and the physical individuality as organisms or as psychological consciousnesses, but instead to seek what is schematically common to the molecule, the organism, and consciousness. In all these cases, the common schema is a domain of absolute survey and activity. Neomaterialism is the result of the survival of thought's old habits in the interpretation of science's new data.

16

Neo-Darwinism and Natural Selection

Independently of every Darwinism, old and new, it is always possible to "explain" in the abstract any fact of finality by invoking chance and the exhaustion of fortuitous combinations. Even if the fact so explained is as improbable as a deviation from Carnot's theorem and as the spontaneous freezing of boiling water, one can always invoke the immensity of time. However enormous the denominator of the fraction that expresses the phenomenon's unique chance may be, it can always become insignificant relative to a more enormous number of centuries. From "Democriticians"[1] to Abel Rey, we constantly find this argument in its pure state. It leads us to believe in the indefinite repetition of the improbable, in a supposedly infinite space-time, in the plurality of similar worlds, and in eternal return. It has no other import than stating the well-known truth: "The series of numbers is infinite." It obliterates every reason and every science. It permits neither deduction nor induction. It does not allow us to distinguish a phenomenon and a miracle. It mistakenly postulates the infinity of space and of time and the eternity of the particles with which chance plays. As soon as we specify the conditions of application of "fortuitous combinations," we realize that the power of chance is very limited.

Suppose we ask someone this question:[2] "Consider 1 million planets, each inhabited by 2 billion humans. Each of these humans ($10^6 \times 2 \times 10^9$) during 1 billion years, tosses every day a die forty thousand times (in one thousand series of forty), that is, practically does nothing else. Approximately how many times would a series of forty sixes arise?" The impression is that such a series will be produced at least some of the time. We can wager 19 against 1 that it will never be produced, because ($10^6 \times 2 \times 10^9$) × ($10^9 \times 365 \times 10^3$) is still 20 times smaller than 6^{40}. Because the duration of life on earth is approximately 2 billion years, it is easy to see why it is extravagant to attribute to chance alone the formation of a nervous system, a circulatory system, the eye or the internal ear, whose ordered complexity has no common measure with the arrangement of a series of forty sixes.

Darwinism is obviously a whole other story. It appeals to concrete facts or to what it deems to be facts: the tendency of organisms to grow geometrically; the approximately stationary character of the number of individuals in each species; the struggle for existence; the slight and spontaneous variations of the individuals within a species; differential mortality; hereditary transmission of variations.

Neo-Darwinism accepts some of these facts. But it draws a distinction between heritable or nonheritable variations (or phenotypic modifications) and mutations. Darwin did not doubt the law of genetics, and neo-Darwinism has practically become an application of genetics to the problem of evolution.

So there is theoretically no intrinsic relation between concrete doctrines like Darwinism or neo-Darwinism and the abstract Democritean reasoning. We do not see how the aforecited biological facts resemble a "spattering in all directions" of atoms and could lead to conclusions that abolish every finalist interpretation. On the contrary, they have sense only on the basis of a properly and specifically biological reality that has a biological sense. Darwin presupposes the tendency of organisms to persevere, to grow, to adapt (he was Lamarckian in this respect). Neo-Darwinism makes use of genetics, it reasons on the basis of genetics, which, if we are not mistaken, bears on a set of biological and even particularly subtle and complicated facts. Darlington, a neo-Darwinian, tried to reconstitute the very evolution that led to the ordinary genetic system of plants and higher animals with diploidy and meiosis, a system on which the neo-Darwinians base their reasoning.[3] Yet he clearly refers back to biological facts, to organs with roles and functions and not to the fortuitous play of molecules. Darwinism and neo-Darwinism are biological theories; they lose their meaning if they lead to the abolition of biological facts as such in order to slip into pure "Democriteanism."

And yet there is hardly any doubt that, mistaking the true content of their theses, Darwinians and neo-Darwinians perpetually conflate, on the philosophical plane, those theses with the old Democritean idea. What accounts for Darwinism's popular success is precisely that it seems to eliminate finalism, the mind, that it emancipates the mind from the onus of believing in the mind; that it appears to reconcile the observed de facto finality with mechanistic and deterministic explanations. It authorized, or so it was thought, an integral monism based on materialism or at least on antifinalism. For thinkers like Th. Huxley, Spalding, and Haeckel, and in particular for their successors, "Darwinism ceased to be a tentative scientific theory and became a philosophy, almost a religion."[4] By the same token, the neo-Darwinism of R. A. Fischer,[5] Th. Morgan, Dobzhansky, Sewall, Wright, Darlington, Julian Huxley, and Simpson rests no doubt on extremely precise and careful studies, but philosophically it has the same Democritean resonance as its ancestor. It is considered in the eyes of its supporters as a way of interpreting the facts of finality without resorting to finalism, a way of accepting these facts while keeping a scientific good conscience.

Thanks to this good conscience, to this confidence in their ability to explain everything mechanically, neo-Darwinians eagerly recognize de facto finality. In his preface to Cott's book, J. Huxley expresses his satisfaction at seeing Cott's demonstration of the genuinely useful and adaptive character of mimetism and animal camouflage. Just as Darwin is often as finalist as Bernardin de Saint-Pierre, precisely because natural selection exempts him from every "overnaturalism," so J. Huxley writes very significantly (in a paragraph titled "The Omnipresence of Adaptation"): "It has been for some years the fashion among certain schools of biological thought to decry the study or even to deny the fact of adaptation.[6] Its alleged teleological flavour is supposed to debar it from orthodox scientific consideration, and its study is assumed to prevent the biologist from paying attention to his proper business of mechanistic analysis. Both of these strictures are unjustified. It was one of the great merits of Darwin himself to show that the purposiveness of organic structure and function was apparent only. The teleology of adaptation is a pseudo-teleology, capable of being accounted for on good mechanistic principles, without the intervention of purpose, conscious or subconscious, either on the part of the organism or of any outside power."[7]

This attitude of neo-Darwinism will dictate our own critique. Its scientific merits are manifest, but we should stress that its Democritean claims are unjustified. Some preliminary remarks are necessary.

a. By recognizing de facto finality, neo-Darwinians dismiss every *divine and vitalistic guidance* as a real scientific impossibility or even as a logical impossibility.[8] Natural selection remains, then, "unless we confess total ignorance and abandon for the time any attempts at explanation."[9] Thus, even prior to every experimental proof, the selectionist theory is favored and the finalist interpretation of the facts of finality is dismissed. If a logical a priori has to intervene here, and if it is necessary to oppose one bias to another, it is on the contrary the finalist interpretation that should be favored, because *in any case,* human activity, which cannot be completely divorced from organic activity, logically requires the finalist interpretation. The theory of natural selection can then push finalism "into a corner," but it cannot reduce it completely. And so we wonder what is at stake in this operation. When we consider that the whole of physics had to be refashioned from top to bottom in the wake of the negative result of Michelson's experiment, whose interferometer could detect a difference in the speed of light on the order of 1/100,000, that it had to be completely refashioned once again in the wake of Planck's experiments on the radiance of the dark body, which led him to define a constant on the order of 10^{-27} erg-second, we cannot help but find childish the scientific politics of biologists who imagine that they do not have to turn the mechanistic frames of science upside down to accommodate human finalist activity and that, in the meantime, they can leave this undisputed finality in its closet. The experience of physics tends to predict that the rejected stone becomes a cornerstone.

b. By critiquing neo-Darwinism (or, rather, the Democritean philosophy it espouses), we are not denying the certain, experimentally observed role of natural selection, either in the internal equilibrium of species or in the equilibrium of fauna and flora, or even indirectly in the evolution of species. It is indeed natural selection that, in the human species, condemns to death under our eyes so many "primitive" races, that on the other hand dooms to an impending extinction so many species of large mammals that we try to rescue in legal reservations. But a favorable or unfavorable action is one thing; a power of organic formation, which would do away with every finalist direction, is another. *We often believe that we are carrying out an experiment of natural selection, when we are simply performing an experiment on the functional value of a given organ.* Experiments on the functional value of camouflage and animal mimetism offer a good example of this confusion. Di Cesnola's experiments on the differential mortality of praying mantis, laid out against a homochromatic or heterochromatic background; Ively's experiments on grasshoppers; and the similar experiments of Carrick, Young, Sumner, and Popham bear on the effective value of camouflage for protection against predators. They are conclusive experiments against Rabaud's thesis, which denies the cryptic colorations of predators every functional value; they are, if you want, experiments on the eliminative or balancing role of natural selection. By definition, a camouflage (or, in general, an effective organ) produces a supplementary chance of survival or a differential mortality; we can thus check the functional effectiveness only by checking the difference. But these experiments do not bear on the organoformative role of natural selection. Or then we are just playing on the word "experiment." Neo-Darwinians have, of course, a theory, and even, since Fisher, a mathematical theory of the passage from one role to the other, but it is not a theory that experiments can verify.

Between two opponents in a war, the one who creates a new weapon immediately gains a great tactical advantage. By the same token, between two industrial competitors, the one with the superior model takes over the market (although it rarely takes over the entire market, because the inferior model almost always better suits a certain category of clients). But no one has the idea of attributing to Competition or War, considered as entities distinct from the conscious efforts of real individuals in struggle, the formation of the battle tank or the automobile, even when the superior model completely eliminates the predecessor. It is equally imprudent to attribute agency to Selection. Consider a hundred men of the same age, among whom fifty have a heart condition. It is obvious that a differential mortality will affect more quickly the group of heart patients. But it would be a mistake to conclude that Death is the agent that builds the complicated system of heart valves. Even when experiments (like those of Quayle on citrus cankers) lead us to attribute to selection the emergence, in a given species, of a variety that is more resistant to a chemical agent or to a virus, it is not easy to prove that the resistant organisms were entirely passive and that selection did not simply validate the organisms' initiative, just as the military or commercial victory can validate a fortuitous invention. The formation of neutralizing antibodies in the presence of microbial toxins or of foreign proteins is, moreover, a fact of experience.

c. Relative to primitive Darwinism, neo-Darwinism's greatest merit lies in its insistence on the complexity of evolution. As long as we examine the facts closely, especially those that touch on the mechanisms of genetics, we will realize that it is impossible not to distinguish between modes of evolution. Higher animals and plants; animals that reproduce sexually and parthenogenetic or hermaphrodite plants; cross-fertilized and non-cross-fertilized plants do not evolve in the same way. But we have to make the differences intervene in the geographic and ecological situation of species or in the size of their population. Neo-Darwinians no longer speak of the origin but rather of the origins of species. There is certainly no *single* correct theory of biological evolution, any more than there is a *single* correct philosophy of human history. On the contrary, there is no reason for the history of proboscises not to be as different from the history of primates as the history of England is different from that of China. The history of genera and species is truly a history in the strong sense of the term, in other words: an inextricable mixture of chances—fortunate or unfortunate, internal (mutations) or external (variations in climate, segregation, etc.)—and of fortunate or unfortunate uses of these chances by the species or the genus at issue. By critiquing the theory of natural selection, we are thus not replacing one theory with another, an antifinalist unilateral theory with a finalist unilateral theory. Instead, it is a question of introducing at least one factor that has a finalist direction, which operates according to very diverse modes and which has to be combined with the other factors highlighted by the neo-Darwinians or the antifinalist biologists.

If it is true, as Cournot showed, that every "history" in the general sense of the term is characterized by a dualist combination of chance and skill, hazards and "reason," then the historical pace of specific evolutions is a sign of this combination of finalist and nonfinalist factors. The complexity of the history of species, the very fact that it is a history, certainly rules out the idea of an all-powerful finalist guidance, of a biological providentialism in the pure state. There is no possible Discourse à la Boussuet on the universal history of species. Yet the coherent, harmonic, and interesting character of this history also rules out the possibility of reducing it to a series of chances, which no finalist or rational factor would integrate. The history of species resembles neither incoherent annals nor an ambitious discourse on the philosophy of human history. It is remarkable that a neo-Darwinian like J. Huxley could write sentences as "synthesizing" as the following one on the perspective of evolution in general: "Evolution may be regarded as the process by which the utilization of the earth's resources by living matter is rendered progressively more efficient."[10] He even speaks of "methods" by which living beings carry out this exploitation of the Earth, as well of the general "progress" in biological evolution. The same statement could be invoked with respect to human history. Despite their innumerable errors, and despite the chances and accidents, humans also "exploit the resources of the Earth more and more efficiently." Neo-Darwinians would have to explain by what miracle these two "histories" can resemble each other if the one (the biological history) only depends on blind mutations without any component of guided invention, whereas the other (the human history) is incontestably "dualist."

Having said that, we can move to the critique of neo-Darwinism proper. One of the crucial factors in the resurgence of the theory of natural selection seems to have been the mathematical study of R. A. Fisher and S. Wright on the probable time a dominant or recessive mutation takes to spread to a substantial portion of a given population. R. A. Fisher, S. Wright, and J. B. S. Haldane thought the order of magnitude of this time was in good agreement with the order of magnitude of species evolutions according to paleontology. Aside from a small number of species in which the rate of mutations is abnormally high, the mutations are produced at the average rate of one mutation per one hundred thousand individuals. This mutation—considered dominant if it provides a selective advantage of one per one thousand, in other words, if the carriers of the mutation have one supplementary chance over a thousand to reproduce relative to nonmutants—will take around five thousand generations to be established in half the individuals of the species and around twelve thousand generations more to be established in the entire species. For a greater selective advantage, the number of necessary generations naturally declines in inverse proportions. The numbers, of course, differ for a recessive mutation. They also vary—and this is the key point that the calculations foreground—according to the numerical size of the population in question or according to the segregation it undergoes in occupying extended areas. For moderately abundant species, like the various species of equidae, the numbers correspond approximately to what paleontology observes, where around one hundred thousand generations are necessary for an evolution that can be considered to be a passage to another species.[11]

But the way in which the numbers are introduced in this kind of problem is necessarily very arbitrary.

Thus in their fundamental formula, Fisher and Wright use a term that characterizes the degree of fitness *of a race or a mutant variety relative to another race or to nonmutants. They stress that this "fitness," or* Eignung, *is not a vague "adaptation," or* Anpassung; *it is the selective advantage, quantifiable as the probability of reproduction relative to the probability of reproduction of another race or of nonmutants. The formula assumes that this degree of fitness of the mutation remains constant during the enormous duration it requires to become fixed. We start to doubt this hypothesis when we reflect on the innumerable variations in humidity, temperature, insolation, food abundance, possible infections, virulence of predators . . . , that can incessantly modify such a term and even transform an advantage into a drawback.*[12]

We are touching here on the fallacious character of mathematical formulas in such a subject. The introduction of a vague event (like the degree of fitness) into the formula requires a stunningly precise delimitation of the concept. Instead of a philosophical concept, we have an exact fraction or an exact relation between two fractions. Unfortunately, the vagueness disappears only to be transformed into a patent falsehood: a mutation cannot give mutants a constant differential mortality during thousands of centuries. If we were to introduce new terms into the formula to represent possible variations, these terms would always be grossly insufficient. If we were to claim to represent with a formula the rate of differential growth of the French population and the English population throughout history, we would have to complicate the formula a great deal; it is doubtful that we could capture the facts, even remotely. And yet, as we have seen, the history of species is indeed a history in the strong sense.

Norbert Wiener made a critical remark concerning social statistics that applies perfectly to the formulas of neo-Darwinians. A good statistic requires extended observations but under essentially constant conditions, just as a good "resolution" of light requires a large-aperture lens. But the lens should also be made up of homogeneous material; otherwise, the aperture would not increase the resolving power. This is why the advantage of long-range statistics under variable conditions *is "specious and spurious."*[13] *A fortiori, one might add, if this "long range" is only obtained by extrapolation.*

It is easy to imagine ways of calculating the chances of an origin-by-selection of a given trait, which would lead to different results. Let us take the case of

Edalorhina buckleyi*'s camouflage bands, which are connected from the thigh to the segment between knee and heel, and then to the segment of the tarsal region. As Cott notes, not only do the intervals between clear and dark bands have to correspond but the sequence of bands in the intermediary series has to be inverted.*[14] *If the bands are marked in their apparent order on the bent leg (Figure 43) as ABC . . .* abc . . . αβγ . . . , *the automatic order in the extended leg will be the following: ABC . . .* cba . . . αβγ . . . *In short, the bands have to be in a well-determined order. If we want to explain their origin by fortuitous mutations and selection, we face a mathematical problem of combinations where the factorial of the number of elements intervenes. Let us even reduce the elements that have to be combined to 10; let us further admit that the first series is selected by chance, and let us only take into account the order and not the spatial connection. For the third series, chance is then the factorial of 10 squared,* $(10!)^2 = 1.3 \times 10^{12}$*; that is, one chance over 1,300 billion. Origin by mutation and selection is mathematically ruled out altogether. For if we admit no direction in the mutation, and if we do not arbitrarily presuppose a single mutation that immediately provokes a mimetic pattern, in one stroke and without explanation; if we instead presuppose small and arbitrary mutations, each of which is selected, then we have to multiply this enormous number by* 10^4 *generations. And for this to be true, we also have to postulate that the small supplementary details in the right direction produce a superior relative fitness of 1/1,000 for each given favorable mutation. We are far from the few hundreds or even tens of mutations that Wright and Fisher estimate can move us from one species to the other in the lineage of horses. The "large scales" of time that are the most implausible even for astronomers and physicists would be widely off the mark. With three rows of fifteen elements each (which is still below the true number), the probability of coincidence for the third row (calculated by the simplified Stirling's formula) is 1 chance over* 1.7×10^{24}*: fewer than 1 chance over 1 million billion billion (always to be multiplied by* 10^4 *generations).*

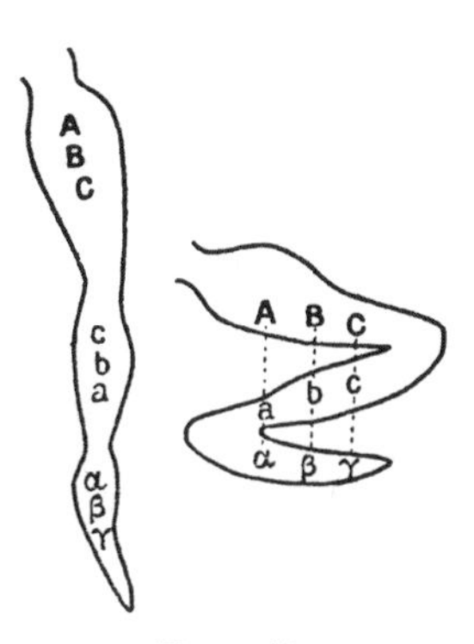

Figure 43.

We will find that these numbers mean nothing. This is indeed our opinion. But we fear that Fisher's calculations may be equally meaningless.

On the other hand, the calculations that deal with the number of generations necessary to fix a favorable mutation in one species (by selection) postulate that the species does not undergo at the same time unfavorable mutations tied to the favorable one. But in fact a mutation that actually or virtually constitutes a preadaptation (a mimetic preadaptation, for example) is often tied to a general weakening of vitality in such a way that the selection eliminates the mutants instead of favoring them.

The individuals of a species cannot be compared to blind beings who have to travel from a *to* b *on a surface that has become deadly everywhere, except on the narrow path indicated by the solid line (Figure 44). Darwinians, old and new, have in mind a schema of this kind, and it seems natural to them that at the price of a sufficient massacre, some survivors will arrive at* a'*, then at* a''*, then at* b. *There is indeed a lottery winner, while for each new selected number, thousands of ticket holders are eliminated. But this schema of the unique path is altogether misleading. By what miracle could the thousands of mutations necessary for the construction of a slightly complex organ calmly follow one another so as to simulate an orthogenesis in an organism that, who knows why, would be protected against every lethal or unfavorable mutation? In reality, the lethal mutations represent around one-third of total mutations and the unfavorable mutations more than half of the remaining two-thirds. The true schema of selection must have multiple paths, and the same organism has to move at the same time on* b, b', b''*, and so on. So, to calculate the chance of success, we have to multiply fractions whose denominators are already enormous relative to their numerators.*

To take a concrete example: neo-Darwinians attribute to mutations followed by selection the mimetic color of cuckoo eggs. Especially when the species parasitized by the cuckoo is unique in an extended area, the eggs of the parasite cuckoo exhibit a high degree of resemblance to the eggs of the host bird, a resemblance that bears not only on the color but also on the pattern of the shell. If we want to quantify the probability of the mutations required by this mimetism, we have to take into account the fact that they had to leave intact the organs of the bird developing inside this mimetic egg. In every case of camouflage and mimetism, the situation is similar. One of the most remarkable traits of animal camouflage is, as we have seen, the complete independence of the camouflaging patterns from the profound anatomy of the underlying organs. The mutations that had to completely alter the appearance of organs thus had to leave intact the profound anatomy of these same organs. To produce, for example, the appearance of a continuous band of uniform color in the fish Lepidosteus platystomus, *they had to bear on very different organs according to necessarily very diverse procedures. In general, the mutations of*

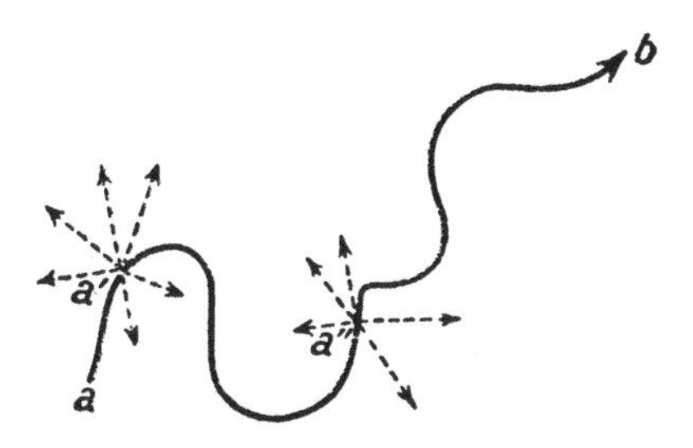

Figure 44.

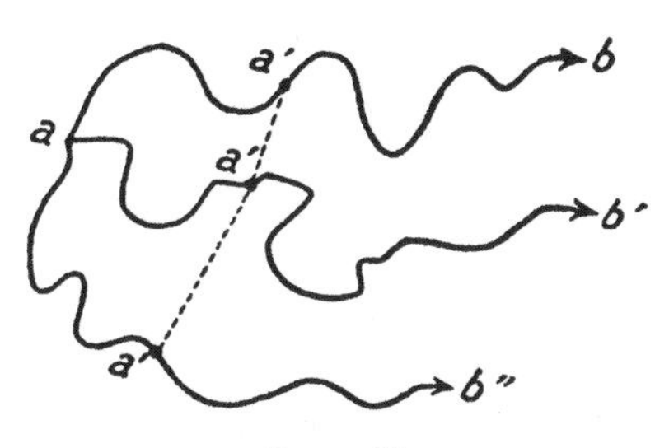

Figure 45.

the laboratory bear on an organ that they render vestigial or whose color they globally modify. Think of the improbability of a series of mutations that can modify one fin or one iris in such a precise way that the fin or iris appears cut in two or into several fragments by a contrast of colors, which extends a contrast obtained in a different way *on the neighboring organ. This reasoning does not apply only to the cases of mimetism. All the organs that Cuénot and A. Tétry dub "tools in living beings" implicate similar convergences. The most diverse organs, with the most diverse procedures, in the most diverse species, arrive at the same arrangements that stem from the very nature and necessities of the operation of the tool at issue. The mutations presumed to lie at the origin of organic tools must then have touched a plurality of anatomic elements in a way that is strangely precise spatially and in a perfect synchrony, while leaving their profound anatomy and their physiology intact.*

To be sure, neo-Darwinians can always displace the difficulty by supposing that a single mutation can have not only multiple effects, which is often the case experimentally, but multiple coordinated *effects. The effect of a single gene in* Primula sinensis *is to incise petals, to double the number of sepals, to modify the bracts, to emboss the leaves, and so on. Why, then, in the case of leaf insects, E. M. Stephenson asks*[15] *after J. B. S. Haldane, couldn't a mutation that affects a single gene produce at once, in a single stroke, several traits that collectively make the insect resemble the leaf? Why not, in effect? So, instead of a series of very improbable mutations controlling the camouflage bands of amphibians, neo-Darwinians can always posit a single mutation that produces all of a sudden these well-coordinated bands. If, instead of ten or fifteen elements in three series, it were a matter of explaining the "eyes" of the tails of peacocks, which are produced by the coordination of millions of barbs of feather, or the patterns engendered by the millions of scales of a butterfly wing, neo-Darwinians would hardly have any choice, because the quantification of probability would lead to superastronomical values. But the difficulty would be merely displaced and not resolved, because this unique mutation would then be the waving of a magic wand; it would not play the role of a natural cause but of a deus ex machina. It would mark a return to the pseudo-selectionism of Empedocles or Lucretius, for whom selection has only to choose between more or less successful or monstrous organisms magically engendered by the Earth:*

> Multaque tum tellus etiam portenta creare
> Conatast, mira facie, membrisque coorta.

Neo-Darwinians took great pains to explain the *coordination* of mutations, temporal as much as spatial. This point is crucial, because paleontology imposes the fact of orthogenesis (the fact of coordinated mutations) and because anatomy imposes the fact of the coherent and adaptive assemblage of organs. According to neo-Darwinism, orthogenesis is superficial; it is not due to an internal tendency, it is reduced to an "orthoselection" (a term proposed by L. Plate) or to "consequential" evolutions (in which a first change produces a "rail" effect and entails subsequent changes). We have thus, as Goldschmidt says, an "orthogenesis without Lamarckism and without mysticism." Consider, for example, how T. H. Morgan argues: "Whenever a variation in a new direction becomes established the chance of further advance in the same direction is increased. An increase in the number of individuals possessing a particular character has an influence on the future course of evolution, not because the new type is more likely to mutate again in the same direction, but because a mutation in the same direction has a better chance of producing a further advance since all individuals are now on a higher level than before. When, for example, elephants had trunks less than a foot long, the chance of getting trunks more than one foot long would be in proportion to the length of the trunks already present and the number of individuals in which such a character might appear."[16] Moreover, the speed of evolution and even its coordination with the evolution of other organs are regulated by the same principle. For a very rapid progression is often not advantageous: an overly long trunk for a torso that is not yet sufficiently massive to a corresponding degree would be more detrimental than useful.

Always in order to explain the coordination of mutation, neo-Darwinians also resort to (1) what Beer calls "clandestine evolution,"[17] in other words, an evolution that bears on the larval or embryonic states in particular, then appears all of a sudden in the adult (by neoteny or fetalization), or (2) an evolution that is not due to the mutation of a unique gene but rather to a complex of small mutations, for the most part recessive and individually unfavorable yet constituting a favorable combination as soon as a final change or a final adjustment intervenes in the genetic complex (R. A. Fisher and E. B. Ford).

These phenomena probably have some basis in reality, but it is not obvious how they can improve the position of neo-Darwinism—if one refrains from secretly introducing a finalist sense into them. If clandestine evolution or if the adjustment of the genetic complex results from chance, or at least according to laws that have no relation to the needs of the adult organism, how is the burden of evolution diminished? Whether we buy a whole ticket or one-tenth of a ticket, we do not alter the advantageous or disadvantageous character of this lottery. As to Morgan's argument, it merely posits the very schema of the theory of selection, while insisting on the supplementary necessity of going at a certain speed from *a* to *b,* and above all of coordinating the progress at the same time toward *b*', *b*", and so on. This argument would improve the position of the thesis only if we forget that it is a matter of fortuitous mutations. J. Huxley, attempting to come to the aid of Morgan's argumentation, makes use of this comparison: "In the evolution of the motor-car, the substitution of four for one or two cylinders was a great progress; it had 'a survival value.' However, not until the majority of cars came to be four-cylinder was the additional advantage of still more cylinders of sufficient appeal to give the six- or eight-cylindered engine any considerable advantage in the market."[18] But from the standpoint of the number of cylinders, the "orthogenesis" of car motors is finalist. Admittedly, the argument is not helpful for proving that the orthogenesis of the elephant trunk is just an ortho-selection. If we recall, moreover, that a mutation in the laboratory is in most cases produced by means as brutal as an X-ray bombardment and that, in nature, equally brutal physical actions are probably at the origin of many mutations, we will be less convinced of the pertinence of the comparison: it is not by bombarding the machinery of a car factory that we would have a serious chance of moving from the four-cylinder motor to the six- or eight-cylinder motor, even with the help of the selection determined by the clients' choice.

Orthogeneses like those of the horse and the elephant are exceptionally favorable to the neo-Darwinian thesis because, at all their phases, they can be deemed to provide an advantage to the animal. But there are harmful or at least superficial orthogeneses, like that of the horns of the Brontotheriidae. How could neo-Darwinism understand an orthogenesis like the one that had to intervene to transform a terrestrial mammal into Cetacea or Chiroptera? Here, the constant general direction of development resembles the general direction of a behavior: it implicates "detours of realization"; we do not see how they can be selected at all their phases, some of which had to be temporarily disadvantageous.

A multitude of observations and even experiments suggest that mutation, far from being the sole material of selection, far from being the elementary rubble out of which evolution is made, is an instrument of plasticity used by the organism. The organism can eventually struggle against a detrimental mutation through modification buffers in the rest of the genetic system (Mather) or through auxiliary mutations that displace the harmful mutations from the dominant state to the recessive state (R. A. Fisher) or, just as easily, let us add, through unknown procedures that perhaps have nothing to do with the genetic system. The bad genes, rendered recessive and harmless by appropriate genetic rebalancing, can become dominant and detrimental again when a crossbreeding destroys this equilibrium. Thus the St. Bernard[19] and the bulldog are races that an artificial selection has pushed to the edges of the normal and the pathological. The St. Bernard simulates acromegaly; the bulldog is almost unviable because of the effect of genes that perturb the activity of the thyroid. The crossbreeding of St. Bernards and Great Danes generates a high proportion of unhealthy individuals (hydrocephalus, paralysis, true acromegaly). The St. Bernard thus lives *despite* the selected mutations that produced it. In the polypoid (multiplication not by two but by three, four, or more of the *n* number of chromosomes), gigantism is a constant trait of the first individuals formed in this way. But this gigantism is very frequently reduced and abolished during evolution, because even octoploid forms are identical in appearance to the diploid form.[20]

We can thus describe as extravagant the thesis—more philosophical (in the bad sense of the term) than biological—according to which selection fabricates and creates all the complex organs of living beings. No known fact even remotely justifies the attribution of such a role to natural selection. Darwinism, old and new, would do well to dissociate itself explicitly from this bad metaphysics that—insulting the memory of Democritus—we termed "Democritean." It is true that neo-Darwinism would then lose a good deal of its prestige: it would have to believe in finality.

Natural selection is akin to competition and war, which stimulate inventions and technical advances, which synchronize the means of attack and the means of defense, which at times eliminate the uninventive individuals or peoples, or which more often reduce the vanquished to a modest "ecological niche." By themselves, they create nothing. Rewarding the inventor is never synonymous with inventing. The direct role of selection is more restricted. It seems capable of establishing gradients of traits (height, size relative to various parts of the body, pigmentation, etc.) in species with an extended geographic habitat, when optima of these traits exist for a temperature or a given humidity. But in general it is rather conservative, either of the average of a species or of the equilibrium of fauna and flora. It ordinarily eliminates the extreme individuals and favors the average type. When a species or a genus has a vast place free of competition at its disposal (fish in the Great Lakes without predators; marsupials without competing mammals in Australia; edentate of South America or insectivores during certain favorable ages; birds in Hawai'i and the Galápagos Islands), the *absence* of selection or the reduction in the pressure of selection enables a diffusion of the favored genus, which can engender a whole fauna without "specialists" coming from foreign orders (the Australian marsupials have "moles," "wolves," etc.).

Their own examination of the facts forced neo-Darwinians (notably Sewall, Wright, and Simpson) to oppose, in many cases, the pressure of mutation or evolution and the pressure of selection *as two antagonistic forces*. If the diffusion is produced particularly in the absence of selection, then it is presumably the pressure of selection that prevents it, at least when the species is in equilibrium with a stable environment. By the same token, a mutation that is unfavorable to a significant population is held in check by the pressure of selection, which only allows a limited field to it. Selection acts in the manner of the force exerted on a semipermeable membrane, balancing an osmotic pressure (the "pressure of mutation" is in this case analogous to the osmotic pressure). The equilibrium is interrupted when the conditions of the environment change or when the size of the population is markedly increased or reduced. A *drift* then takes place, which can lead either to the complete elimination or to the total generalization of the mutant gene. This equilibrium is easily disrupted in small populations where accidents and chance play a substantial role relative to statistical laws, especially when the pressure of selection declines at the same time.

Selection thus embroiders minor variations of details on the major, truly creative variations of the fundamental and effective organs and systems that account for the success of the major dominant types: sexuality, meiotic system, internal fertilization, homeothermy and homeostatic mechanisms in general, aerial respiration, formation of feathers and wings, centralized nervous system, and so on. But we have not observed the emergence of a new organ by neo-Darwinian factors (mutation and selection) any more than we have brought off the impregnable experiment of Lamarckian heredity of acquired traits. Much less publicity has been made over the first of these two negative results. It is fair to reestablish the balance. We can provoke mutations in the laboratory very easily and even, as we have seen, proportionately to the dose of X-rays used, but nothing allows us to consider these mutations as elements in the construction of a new organ.

17

Neo-Darwinism and Genetics

We have yet to present the most decisive argument against both neo-Darwinism and Schrödinger's neo-materialism, against the thesis that tries to explain the formation and evolution of species and the de facto finality of organisms through genetics and through a mechanical selection of fortuitous mutations.[1] This argument can be drawn from the facts uncovered by experimental embryology. These facts[2] indisputably testify against the theory of the mosaic-egg, because even in the young gastrula, grafts transplanted quite precociously can develop *ortsgemäss*, according to their new localization, and not *herkunftgemäss*, according to their origin. For example, a ventral fragment of the ectoderm, transplanted on a branchial region, develops branchial fenestra at this site. To avoid the recourse to agents of finalist and transspatial regulation, to a true epigenesis of embryonic structures—an epigenesis that biologists find profoundly repugnant—the embryologists have only one choice: to imagine preformation, the "mosaic," in the genes. The transplanted graft develops *ortsgemäss* because the same specific genes, according to different inductive influences which stem, for example, from a certain level of one or several inductive substances, will supply ventral, neural, or renal tissue or branchial fenestra, and so on. A graft of frog, of triturus *cristatus* or *taeniatus*, of axolotl, will only ever produce tissue of frog, triturus, or axolotl; but according to the place where it is inserted in the host (whether this host is frog or triturus), it will produce ventral skin, or branchia, or a kidney.

But it is easy to see the crushing burden imposed on genetics by examining individual development. Because the inductive substance is an ordinary chemical, the whole "responsibility" for structural development is obviously assigned to the system of genes. It is the genetic structure that has to explain the structure of the adult organism. But this "explicative" structure has to be "nested," with multiple overlays, because depending on the induction it will undergo, it has to produce organ *a* or organ *b,* and so forth. The theory of the mosaic had some plausibility when we were dealing with the egg or the young embryo, taken as a whole and experimentally untouched; it no longer has any when it claims to transport the "mosaic" into the genes. These genes cannot explain *at once* the frog-trait, the triturus-trait, or the axolotl-trait, on one hand, and, on the other, the leg-trait, the kidney-trait, or the branchia-trait, or the countless organs that we can enumerate.

It is moreover a widely known fact that genes remain what they are in all the cells of all the organs of the body, because mitoses are genetically equal. They *do not become* the adult structure that they supposedly explain. Geneticists have spent treasures of patience and genius to establish the map of genes in certain species like the Drosophila. They studied the points of chromosomes, the *loci,* which observation and induction reveal in correlation with this or that trait of the adult (e.g., "vestigial wing," "vermillion eye," "bar-eye"). They ought to be excused if, in the enthusiasm of their discovery, they thought they resolved the problem of total heredity, ontogenesis, and phylogenesis.

But it is not in fact clear how a one-to-one correspondence can be established between the structure of genes and the complex structure of adult organs. What is clear is the *modifying* action of genes on a structural formation that is given with its own laws. It is obvious that a gene produces or provokes the formation or, having mutated, alters the speed of formation of a chemical capable of modifying the color of the eye, inhibiting the development of the wing or making it curly, creased, or deltoid. But how could one or several genes remotely control the normal structure of the eye, the wing, or the nervous system of the drosophila? And where are these so-called normal structure genes on the maps of the Morgan school? If the genes harbored the secret of total heredity, the map of chromosomal loci of the drosophila would have to resemble a diagram of the adult drosophila. We would need something like the schematic homunculus that can be established on the precentral gyrus in the wake of experiments of electric stimulations, where the general structure of the human organism can be roughly recognized (despite the different proportions and the fact that the "cortical" tongue is as large as the "cortical" trunk).[3]

Only in an extraordinarily small number of cases is the hypothesis of a *structural* correspondence of genes to the adult organism not implausible a priori. Morgan cites the case of the dextral and sinistral coiling of freshwater mollusks *Lymneae*. In these mollusks, the coiling is normally dextral, but we encounter individuals with sinistral coiling.[4] The study of their crossbreeding indicates that Mendelian heredity is at stake. There exists a dominant dextral gene and a recessive sinistral allelic gene. But—here is the crux—we can track the beginning of the right or left coiling in the embryogenesis of these mollusks from the earliest cleavages of the egg (four or eight cells). It is thus not absurd to suppose "that the characters that develop in the protoplasm are ultimately traceable to the genes in the chromosomes."[5] But these cases seem very particular. The direction of a coiling is a very simple structural character, so simple that it is not in fact a matter of "structure." A "right" glove has the same "structure" as a "left" glove. The example is even less favorable because, in identical "mirror" twins, the normal or reversed situation of organs cannot derive from the structure of genes, because the two individuals have the same genetic structure.

For the least structure, it is true that the *structural* passage gene → egg → embryo → adult is inconceivable (except by magical action).

In reality, the hypothesis should be dismissed, because we have today precise notions about the nature and structure of genes that a whole host of observations (especially, as we have seen, observations with electronic microscopes) show to be similar to the structure of ultraviruses and enzymes. The genes are formed out of nucleoproteins "whose macromolecules are arranged in a definite way along the chromosome, thanks to the latter's permanent skeletal filament."[6] In all the cases where we know, at least in part, how a gene "controls" a trait, it is only ever a matter of a control through hormones, modifying one structure whose law is also given and to which no microstructure corresponds in the responsible gene. The gene that controls albinism in rats "does so because it stops the production of an enzyme necessary for the formation of the dark pigment. A gene that entails dwarfism in mice does so because in the cells of the pituitary gland, it stops the formation of hormones."[7] By the same token, we know how the genes that determine the blue, purple, or red color of flowers act on the chemical reactions that give birth to anthocyanins. From the gene to the hormone whose production it triggers, there is perhaps and even probably a certain structural continuity, but from the hormone to the organic structure, there is certainly none. This becomes obvious when we remember that the gene is supposed to control the emergence of instincts as well as organs. What structural relation could there be between a nucleoprotein and an instinct?

To connect gene chemistry to the adult structure, the last possibility that embryologists (and neo-Darwinian evolutionists) have is to invoke a modifying action that bears on the speed of development of organs and to extrapolate to the point of making the structure of the organism the sum of the controlled differential growths. Theoretically, one can in fact always move from a structure *x* to any structure *y* through modifications in the rate of development of various parts. Like the fanciful etymologists of antiquity or of the seventeenth century, who always found some means of passing from one word to another, it is enough to correctly arrange amplifications and reductions to transform one organ into another, the first as simple, the second as complex as one wishes. By making a rate of differential growth control the amplifications and reductions, hormones control this rate and genes control the hormones in their turn, one has the illusion of explaining a structure by a chemical and indirectly by a gene. But although the sophism of the operation is disguised, it is quite visible: there must necessarily be as many genetic controls as there are details in the structure to be explained for the explanation to be effective. The theory in question is thus just a new avatar of preformationism. Instead of a pure and simple microstructure of the adult organism in the genes, one supposes a microstructure in the system that controls growth rates. This hypothesis is not very economical. On this or that detail, it is perfectly legitimate—and experimentation confirms this thesis—to explain a particular development through a rate of differential growth, through what is called "allometry." Goldschmidt did so for certain intersex forms; Sinnott for certain fruit structures; Swinnerton and D'Arcy Thompson for the forms of mollusk shells; J. Huxley, de Beer, and Lumer for certain facts of tachygenesis or to explain racial traits in some species. But to see in allometry the universal key to the explanation of organic structures, as D'Arcy Thompson tends to do, is to conflate an indeterminate, unlimited theoretical possibility—but as unusable from the scientific perspective as the Democritean mechanical sorting or the fanciful etymologies of Plato or Gilles Ménage—with the bringing into play of precise phonetic or biological laws, which do not dispense with the recourse to a preexisting structure but, on the contrary, presuppose it.

The passage from genetic action to the somatic character (physiological genetics) represents the weak link in the chain of genetics. This weak link prevents it from supporting the weight of embryology or of evolution.

The inevitable conclusion is that, contrary to the hopes of geneticists like T. H. Morgan and embryologists like Dalcq, the two disciplines in no way intersect. In the end, embryologists cannot use the action of genes to understand ontogenesis; they see in genes only modulators or triggers of development whose principles are elsewhere and of an entirely different order. And if genetics is incapable of explaining ontogenesis, it cannot explain phylogenesis. The life of a species is after all just a succession of ontogeneses. This is a truth of pure good sense, completely independent of the well-known theses of Bolk and de Beer.[8] If genes do not explain the normal structure of the organism, then the genetic mutations on their own cannot explain the evolution of this structure.

This problem can mislead us because we can take things from the other end. We can in fact say, by virtue of the numerous experiments on Mendelian genetics, all biologists (whatever their opinion of the scope of genetics) recognize that genes have at least a modifying influence. A mutation creates a new lineage in a species, with recognizable and in principle definite traits. So it suffices to add mutation to mutation to have racial, then specific differences. But the notion of "species" is by general admission very difficult to specify. Alongside the good species (good for the classifier), there are countless others that drive him to despair: polymorphous species, species with geographic gradients, with ecological varieties, and so on. The specific differences are not therefore an impassable barrier. If the basset differs genetically from the greyhound, it is natural to think that the dog differs from the wolf in the same way, and similarly canines from felines, mammals from reptiles, vertebrates from invertebrates. Instead of saying "what cannot explain ontogenesis cannot explain phylogenesis," one can say "what explains the difference between two organisms in a single race can explain the difference between any two organisms."

Logically and in the abstract, this second reasoning is as valid as the first. It is the facts that decide unequivocally against it. In any event, there is one case where the responsible genes cannot explain the structure to which they are correlated: the gene or the heterochromosome that orients toward the male or female sex. It is impossible to consider the sexual structures as the outcome of a series of mutations whose pattern was conserved by the heterochromosome. Goldschmidt's and Witschi's experiments on intersexuality have already shown that the genetic determinism of sex is very relative and that the heterochromosome is just one link in a chain or one factor among others. But the experiments of sex change have decisively demonstrated that the sex genes act only via other hormonal factors, because the same hormones that act on the secondary sexual traits in the adult organism can also trigger sexual differentiation from the beginning of embryonic life, not only in the absence of the genetic sexual determinism but despite the presence of an opposite genetic determinism.[9] No one can maintain that a hormone like androsterone or estrone, with a relatively simple chemical structure, can contain the pattern that corresponds to the structure of sexual organs. Especially because these same hormones can change the sex in the most diverse species and groups and because, conversely, a substance that acts as a male hormone in one group can act as a female hormone in another group. In this case, the hormone acts as a kind of conditional stimulant, and the gene certainly acts in the same way, because it can be held in check by the hormonal action.

What is true for the genes of sexuality is equally true for other genes. They are simple triggers or orientators that do not act directly but via other triggers. Furthermore, there is no need to fall back on theoretical arguments on this point; Baltzer's experiments have shown (in agreement with the experiments of Wolff and Dantchakoff) that the action of a lethal gene or of a mutant gene in a graft can be corrected by the influence of substances emanating from the normal host.[10]

In sum, the facts prove that genes are not microstructures that correspond to the structure of the organism. They are in no way a code-script, as E. Schrödinger says. Strictly speaking, they do not even explain the peculiar structure of the mutant organism, because this structure is very likely the outcome of the organism's active response to the disturbance introduced by the mutant gene.

What do genes represent, then? It seems that the preceding critique is too strong, that it goes too far, and thus disqualifies itself by its very excess. Genes have to serve some purpose. All the experiments of Mendelian heredity, controlled by the genetic system, subsist. But what we have critiqued is the ambition of geneticists to explain with genes, and with genes alone, the structure of organs and the evolution from the structure of one species to that of another. It is the claim of making the gene "the primary organiser and determiner of all structural and functional characters in living organisms."[11] Genes can perfectly have another nature and play another role, as the unprejudiced interpretation of facts suggests. For the sex genes, it is obviously a matter of shunting between two possible paths by a game of heads or tails. It is a game of chance subordinated to a need of the crossbreeding species. The variety of procedures used proves that this game of chance is merely a means. At times, it is the masculine sex that is heterogametic (vertebrates, except birds and lizards; butterflies); at others, it is the haploid or diploid state of chromosomes that serves as "heads" and "tails" (without allosomes); last, heterogametism can be obtained either through a Y heterochromosome (drosophila) or through a single allosome instead of two (this is perhaps the case for humans). And in effect a game of chance "presence of A–absence of A" fulfills its role just as much as a game of chance "presence of A–presence of B."

This commutator or shunter effect, this *switch*[12] effect, is peculiar in its pure form to the sex genes, but it gives a clear indication of what can generally be expected of other genes. Because the sex chromosome simply directs toward the formation of male or female organs without accounting for their structure, it is unlikely that the other chromosomes could have an essentially different and so much more heightened role. A switch effect must necessarily take place in the case of polymorphous species as well.[13]

Certain butterflies, whose females imitate various species that birds find inedible (Papilio dardanus, P. cynorta, P. polytes), *have at times three or four female forms, some nonmimetic, others mimetic and different from one another. The case of polymorphous species presents crushing difficulties for neo-Darwinism. If it were not a matter of individuals of the same species, a neo-Darwinian would attribute the considerable differences in structure and behavior among the females of* P. polytes *to a protracted orthoselection that integrates hundreds of distinct mutations. But these extremely different structures are produced by the same chromosomes (one or two genes generally tied to the sex) that control the switch effect, just like the heterochromosome controls the sex. It is clear that, in polymorphism as in sexuality, genes that* today *condition the orientation of an individual toward a given form cannot be the genes that during phylogenesis were, by their mutations, the origin of the forms in question, as the theory suggests. The genetic study of polymorphous species has produced rather confused results for the* P. dardanus *(studied by Ford).*[14] *For* P. polytes *(studied by Fryer), two factors, A and B, tied to the female sex, trigger one of three forms, depending on whether A, or B, or A and B are dominant. All the males resemble one another, despite a variable genetic constitution, where A and B are indifferently recessive or dominant. Thus there would be a double switch for females; the commutator of the three forms only functions in the presence of the "female sex" commutator. When confronted with such cases, one has to be readily blind to continue to imagine that the mimetic structures can be explained by numerous mutations selected by orthoselection, because in today's species, the presence or absence of infinitely complex details of this or that mimetic structure is conditioned by the presence or absence of one or two genes only. What have the innumerable mutated genes that the theory necessarily presupposes become? If neo-Darwinism prefers to believe that the very origin of the three female forms is due to the same genes (A and B) that today operate the switch, it falls into a magical theory of the role of genes; it escapes finalism only to slip into fairy tales. For the sake of analogy, it will end up considering the sex gene in the same way: the heterochromosome that today determines the individual's sex must be deemed to have originally provoked the emergence of sexual male and female structures—which is no longer even a fairy tale but pure and simple non-sense.*

Outside the cases of determination of sex and of polymorphous forms, the genetic system does not seem to have the role of *switch* toward well-defined forms, and it cannot be described before the toss of the dice that decides for one or the other. But it does always resemble a systematic game of chance, capable of furnishing the species with small preadaptations and thus increasing its plasticity and its chances of survival in an ever-variable environment: tiny differences in dietary requirements; in thermic or hygrometric or light requirements; in the resistance to various infections or deficiencies; in the flight or running capacities. It cannot, of course, be a matter of attributing to genes the emergence of fully preadapted organs; it can only be a question of small, above all quantitative differences in the range of possibilities of a function. In normal circumstances, these small differences do not necessarily constitute a material for a positive or negative selection. They only expand the species's geographic or ecological domain.

The human species should be mentioned with caution, because its life is not purely biological. Consider, nevertheless, the enormous utility that the variety of individual talents and dispositions presents for social life. On another plane, the variety of genetic combinations in an animal or plant species is obviously advantageous for this species. Selection seems to intervene only in exceptional circumstances—drought, famine, epidemics—that considerably reduce the size of a population; it can thus eliminate certain genes from the species, thereby diminishing its plasticity, until mutations remake the lost genes when the circumstances become favorable again.

Genetic theory, neo-Darwinism, and neomaterialism postulate that one and only one organic structure corresponds to every genetic system (not taking into account completely dominated recessive genes); that every variation in the genetic system entails a variation in the organic structure; and that, inversely, every structural variation makes it possible to suppose a prior variation in the genetic system. But this is a postulate and not a demonstrated proposition. Nothing proves that a series of mutations or even any change in the species's chromosomes corresponds to a series of orthogenetic organic forms. Nothing proves that from the eohippus to the horse, genetic mutations commanded the atrophy of lateral fingers. Instead, a good number of indications imply the opposite. We have already cited the case of giant polypoids that progressively return to a normal size. We can add numerous facts verified in the laboratory, in which a provoked mutation, at first unfavorable and diminishing the vitality of mutants, is progressively better and better supported. So the return to normal stature, like the return to vitality, is in fact independent of a new change in the genetic system. To salvage their hypothesis, neo-Darwinians introduce here the auxiliary postulate of genetic "modifiers," which neutralize the mutant gene. But it is at least just as plausible to admit an organic action of a nongenetic nature, as the phenomenon's progressivity suggests. The so-called reverse mutations, whose effect disappears very quickly (like the mutation that controls the miniature wing in drosophila), justify analogous conclusions. But then, if the species were capable of returning to normal *after* a mutation, why would it not be capable of modifying itself *in the absence* of every mutation? The link of a given genetic combination → a given somatic structure may very well be provisional, like the link established in the order of individual psychology between a conditioning stimulus and a response. Obviously, the "provisional" here is on a whole other order of magnitude than the psychological "provisional." But the action of a gene is capable of "extinction" just like the conditional reflex, though at the end of an infinitely longer time. The sex chromosome's mode of action by *switch* has all the traits of a conditioning signal. In any case, it has the trait common to all signals: it has an arbitrary nature.

This is why mutationism had much more success than Lamarckism: its experiments bore on actual effects and on short durations. Every novel genetic combination, every mutation, must produce an immediate effect on the organism, just as in the experiments of conditioning, every change (even minimal) in the stimulus situation is translated by a difference in the behavior of the animal. Yet, just as the salivary function of the dog is itself independent of the arbitrary stimulus that triggers it, the somatic structure is probably independent of the genetic complex with which it is provisionally associated. When we pass from the biology of the laboratory to planetology, it is striking to see the Lamarckian hypotheses regain the advantage. Paleontologists are very rarely neo-Darwinians, and the failure of neo-Lamarckism in experiments of short duration hardly impresses them.

It would probably be interesting to take up from this new perspective and to invert the so-called principle of organic selection formulated by J. M. Baldwin[15] *and Lloyd Morgan. According to this principle, an organism at first adapts to a new environment by a change in habit or a particular direction of instinct without a genetic basis. Then the mutations that emerge and appear suitable—in their effects on the organism and the instincts—to this new life of the species are favored by selection. In sum, what is at stake in this case is a simulated Lamarckism: nonheritable modifications are later fixed in the species by mutation and selection. But if the interpretation of the genetic system as a switch or a "conditioning signal" is true, there must also be more numerous and more significant cases* where it is mutationnism that is simulated: *a gene or a genetic system is progressively tied to an organic structure and to an instinctive behavior that no mutation provoked (as any stimulus can be tied to an instinctive behavior). It appears, with due cause, to be the key to the structure and the behavior, even though it came after them. This principle, while inverting Baldwin's, has something in common with it: it appeals to the opposite of a genetic preadaptation. It appeals like organic selection to a genetic "postadapation," without postulating a mutation. At any rate, whatever the scope of this principle may be for the ordinary genetic traits, it is impossible to explain the genetic control of sex, or the form in polymorphous species, in a different way. Sexuality undoubtedly preexisted the genetic "signal" that determines the male or female sex; and by the same token, the various mimetic structures in a polymorphous species are logically independent of the single gene or of the two or three combined genes that guide the individual toward one or the other of these structures.*

The specific organism can be compared to a sophisticated machine that is capable of accomplishing very varied performances and the genetic system to a modulating keyboard (akin to the play of the harmonium's timbers). This keyboard does not contain the general structure of the machine in reduced form; it does not control its general operation, it is neither an organizer nor a motor, it is only a modulator. An engineer can easily replace this or that organ of the machine while conserving the same modulating keyboard, or vice versa. But for individual usage, every action on the keyboard is translated by a difference in the operation of the machine. But this comparison is inadequate, first of all, because chance and not the individual user controls the genetic keyboard. The dominant traits of the user's temperament are fixed from the moment of his conception by heredity's game of chance. It is again inadequate because the organism is not a machine and the keyboard does not control the mechanical system it triggers, but probably mnemic themes it evokes with "signals." It is inadequate in countless other ways as well. But there is one way in which it is perfectly valid: it is just as absurd to explain the structure or evolution of the organism by the chromosomes as it is to explain the harmonium by the play of timbers or the car by the dashboard. The genetic mechanisms do not exempt the biologist from the recourse to finalist factors; they are organs in the service of a finalist direction.

18

Organicism and the Dynamism of Finality

Organicism is an imprecise term, which has the merit of corresponding by its imprecision to the vagueness of the doctrines it designates. These doctrines have something in common: they claim to escape both determinism and finalism, or they claim to reconcile the two. Organicism neither seeks to reduce the organism to physicochemical phenomena[1] nor seeks to explain organic specificity by a distinct principle (vital principle or soul), which would intervene *dynamically* in the unfolding of physical phenomena. Even if the parts taken individually conform to physical laws, their assemblage, the organization in its totality or its unity, is enough to reveal the specific character of the organism. Several theories of "totality" (Smuts's "holism," *Ganzheitlehre* of Alverdes and of Bertalanffy) can be classified with organicism, because they equally insist on the necessity of considering the organism in its entirety. To see the organism as a whole is the key. Both the organicists and "*Ganzheit* theoreticians" conveniently believe that the problem of interpretation blends with the problem of objective explanation or, rather, that it replaces it.[2]

Organicism and affiliated theses have had the greatest success, especially in Germany. One cannot help but think that one of the reasons for this success is the imprecision of the doctrine. In a difficult problem, but which can always be decided by experimentation, one dare not take a stand. Organicism presents itself then as a third party. Like those political assemblies, "determined to maintain state intervention while promoting liberalism," organicism declares, "recognizing the full validity of physico-chemical laws in the order of life, but considering the organism as an unanalyzable and absolutely specific whole; seeing in organization a factor of unity and of regulation, but avoiding turning this factor into an active and transcendent agent; rejecting every introduction of vitalism or animism." Organicism in its doctrinal aspect has all kinds of advantages; it seems to be scientific and positive, it can appeal to experimentation at the same time as to phenomenology; it can insist on the implausibility of physicochemical reduction, while doing away with the appeal to a "force" that intervenes in the unfolding of physical phenomena and critiquing its metaphysical uses.

Regrettably, this advantageous doctrine has a drawback: it exists in words only. Organicism is an empty concept that has no basis in reality; it is a "squared circle." If an act or a being with a unitary, finalist, and organized appearance can be completely explained by factors that are fully subject to physicochemical laws, then by definition it is not truly unitary, finalist, or organized. It is merely an "aggregate" or a system of equilibria. Conversely, if an act or a being is truly unified and organized, then by definition it cannot be reduced to a set of physical processes that propel or balance one another.

Two clarifications are indispensible here:

1. A manufactured machine, one will say, is unified, finalist, "organized," eventually self-regulating, and yet it obeys a step-by-step physical causality. But as we have seen, a machine is indissociable from the living being who established it. It is an external organ. No one can deny that the organism comprises countless machinic functionings or "substituted chains," substituted for survey, for the finalist factor and for its primary action. But considering the works of organic finality as connected is not tantamount to creating a theory of finality. Terms like *organization* inadvertently have a double sense, active and passive. Once established, an "organization" can function through a step-by-step causality, while responding with its assemblage to the goal pursued by the "organization" in the active sense of the term. The problem is to understand organization in the active sense, because living beings are not ready-made. In biology, the problem of origin and of formation is indissociable from the problem of nature. Organicists like Rostan,[3] the founder of the theory, and Delage eventually discovered that Descartes was their predecessor, because for him "the digestion of food . . . respiration, wakefulness, and sleep . . . naturally proceed from the mere arrangement of organs."[4] The power to live, says Rostan, is not a separate property; it is "the assembled machine."

But the name of this theory is obviously "mechanism" and not "organicism."

2. There is one and only one case where it is conceivable, if not without some implausibility, at least without logical contradiction, that determinism and finalism are simultaneously true: the case of the universe as a whole. All metaphysical monisms can cling to and affirm at once the freedom of the absolute and the reign of necessity in the world. They can believe in the harmony between the reign of mechanism and the reign of divine providence. But in this system, individuals and individual organisms have no genuine reality. This is not what organicism can accept, given its affinities with "holism" and its insistence on the real autonomy of the organism in its unity.

Kant was not a monist, and he shared his century's taste for final causes. Nevertheless, because he believed in the mechanistic and deterministic science of his time, and because his critical position forbade him the slightest doubt about the universal value of determinism, he adopted for the universe of science as a whole a point of view very close to organicism's; he did not really distinguish—despite the famous opposition between internal finality and external finality—between astronomy or geography and biology.[5] And it is clear that Kant's theory strongly influenced subsequent organicists. The mechanistic explanation is universally valid and exhaustive; but the teleological judgment is always just as legitimate, although it is only reflective, because "nature clearly presents a final unity of intention." Kant knows full well that one can meditate piously (like Fénelon) on the harmony of nature and even on the utility of vermin for inciting man to cleanliness[6] or on the utility of dreams "[for moving] the vital organ internally by means of the imagination and its great activity . . . and in the case of an overfilled stomach, where this movement during nocturnal sleep is all the more necessary."[7] In contrast, he does not admit that finality is introduced as a particular cause in the explanation of the formation or behavior of a living organism. The two ends of the chain are joined only in God. Nature, deterministic according to the Understanding, and nature, finalist according to Reason, are harmonized by the faculty of Judgment. But this faculty is "referred to the supersensible" and the unity arises "in an unknown way." The final cause is not a force; it is merely a legitimate and indispensible point of view, not only on living beings but on the entire world.

Kant's thesis had the greatest success throughout the whole nineteenth century and even up to the beginning of the twentieth, until quantum physics and the crisis of determinism.[8]

Organicism applies the "monist" thesis to the scientific study of organisms, without realizing that in this way it loses its meaning. In his biological philosophy, Claude Bernard accumulated contradictory hypotheses:[9] "Everything is derived from the idea [that guides vital evolution], which alone creates and guides."[10] But this "idea" is not effective. "The vital force directs phenomena it does not produce; the physical agents produce phenomena they do not direct."[11] The vital force pertains to the metaphysical world; "great is the error in believing that this metaphysical force is active."[12] "There is an irremediable error at the basis of *vitalist* doctrines, which consists in considering as a force a misleading personification of the arrangement of things."[13] Because Claude Bernard was nevertheless a sensible man, we have to admit that he referred back to a metaphysics like Kant's and that he escaped the contradiction by returning to monism or to the unity of the simultaneously biological and cosmic "initial impetus."[14]

We are reluctant to sketch out a study of contemporary organicists. They accumulate subtleties to conceal the uncertainty of their thought, and they believe that the double negation of mechanism and finalism is equivalent to the affirmation of a novel thesis.

W. E. Ritter[15] is first and foremost a holist, and he comes quite close to admitting that the whole dynamically determines the nature and behavior of parts: "The organism itself as a living whole is a factor in determining the nature of the cellular elements of which it is constituted."[16] "The organism is individualized and unified in such a way as to give it as one whole a measure of determinative power for its own welfare over each of its parts."[17] But Ritter has in mind a Gestalt-form's wholly relative power of determination: "A natural whole stands in such relations to its parts as to make it and its parts mutually constitutive of each other."[18] We have sufficiently insisted on the pseudo-finalist character of Gestalttheorie *not to return to it here. In a Gestalt-form, the power of determination of the whole over its parts results from a law of extremal equilibrium, and we do not see how it could achieve "its own advantage."*

F. Alverdes is closer to a genuine finalism, and he distinguishes the Gestalten *from "organismic" wholes without clearly specifying how these latter are distinct from the others. L. von Bertalanffy provides a perfect sample of organicism's indecision.*[19] *Organicism can in principle be described as a set of physicochemical processes. If we study the vital processes from the viewpoint of the physicist and the chemist, we never discover a process that contradicts physicochemical laws; in this sense, life is merely a combination of physical and chemical processes. And yet this description misses the heart of the matter, namely, the combination, the particular organization of these processes that perform a "function" by virtue of this organization. The same thing, as we have seen, can be said of a machine once it has been built. Up to here, Bertalanffy's thesis seems to be that of finalist mechanists of the seventeenth century. But Bertalanffy accepts the irrefutable character of Driesch's arguments and rejects the mechanistic thesis. He equally rejects the forms of vitalism and of animism. According to him, the organicist interpretation makes it possible to escape the mechanistic doctrine as well as other doctrines, finalism as well as mechanism. Organic finality is only an "as if." Because, unlike Kant or Claude Bernard, Bertalanffy does not seem to contemplate a monistic and divine origin for both physicochemical causality and the apparent causality in the organismic assemblage, it is not clear how he can sidestep the contradiction. He escapes it provisionally by presenting the organicist interpretation as a pure description and not as an explanation: "This interpretation leaves open the question of how organic totality is actually maintained." But our author hastens to contradict himself on this point as well; he admits that for the scientist, the only question is to know which explanatory principles are necessary and sufficient for the vital processes. The organicist interpretation becomes an explanatory principle each time biologists hypothetically use purely biological concepts. And Bertalanffy cites as examples Schaxel's theory (persistence of form), Heidenhain's theory (syntony), and Gurwitch's theory (which, to explain the morphology of mushrooms, appeals to a* field *or* morph *that belongs to the germ itself, influencing the mitoses and cellular growth by imposing an overall form on them). But in the end, Bertalanffy reverts to the thesis of pure interpretation: the various theories are not really "explanatory." We find the same hesitations in Dalcq. From 1935 to 1947, he oscillated between the affirmation of a specific vital activity, a de facto finality, and the purely physicochemical explanation. The title of Dalcq's major work,* The Egg and Its Organizing Dynamism, *is altogether misleading: only gradients of substance and chemicodifferentiations are at issue in the text.*

Bounoure[20] cannot be ranked among the organicists or the holists he critiques; rather he is, it seems, a vitalist, because he insists not only on the autonomy of the organism but on the substantial and transcendent character of the specific principle of this autonomy. Between the order of the mechanism and that of psychological consciousness, life retains for him its proper originality, which resides in its "dual character: organized matter and organizing idea." But how does this idea act dynamically on matter—on a matter that has to be considered substantially distinct? For Bounoure rejects as "romantic"[21] every panpsychism that would attribute an autosubjectivity to molecules and does not appeal, like N. Bohr or Lillie, to the new physics of the individual. On this important point, Bounoure repeats all the hesitations of organicists, and it is typical that he cites Claude Bernard and his "guiding idea" that alone creates and guides. It seems that this reference to Claude Bernard dooms the writer to all the organicist vacillations. The autonomous power of regulation, which distinguished the living being from a clock, does not preclude "the general justification that modern biology brought to the doctrine of physico-chemical determinism."[22] However, material determinism is nothing more than the necessary auxiliary of the organic form. Consciousness cannot be invoked "as a binding and modeling force," because this invocation would "resuscitate vitalism." Contextually, Bounoure understands by "vitalism" what is generally designated as "animism," and he does not condemn every vital dynamism. But he imagines it as transcendent in the theological sense of the term and not only as transcendent relative to physicochemical processes. The specificity of life thus stems directly from God: life contains something unknowable and miraculous. He does not specify whether God intervenes atemporally, through preestablished harmony, or through active temporal influence. It is not easy to discuss a theological thesis; nevertheless, even at the price of this admission of scientific impotence, the author does not escape contradiction: if vital specificity stems from God, in what sense can we speak of the "autonomy of the living being"?

Exactly parallel to organicist biology, there exists an organicist psychology that strives to reconcile determinism and finalism or to reject both for the benefit of an "organismic" point of view. In his rich work, K. Goldstein[23] critiques with particular vigor the "analytical" explanations, which artificially isolate elements such as the reflex, the conditional reflex, to reconstruct the psychoorganic behavior piece by piece.[24] He opposes to them the point of view of "organic totality."[25] Every vital process presents a ganzheitliche Gestaltung, *which connects it to the temporary situation of the rest of the organism; this organism realizes compensatory adjustments and displacements, which are meaningful substitutions. A determined performance does not depend on the functioning of a given region.*[26] *After the transplantation of nerves, the performance can nonetheless be equally successful without exercise. The form of incitation does not depend on a given anatomic structure.*[27] *Even the perception of a color exerts an action on the entire organism.*[28] *It is against the backdrop of the entire organism that the form of a behavior or a privileged perception,* ausgezeichnete Verhalten, *stands out. Goldstein supplies even more arguments than Lashley does, not only in favor of cerebral but also in favor of organic equipotentiality and of the interpretation of the organism as a domain of absolute survey. But he does not provide, in contrast, a precise and positive response to the problem of totality's mode of action. Like organicists, he admits that because his point of view does not reduce everything to mechanical processes, it also does away with the hypothesis of entelechy, either in the form of N. Weyl's and Riezler's* Material Agens-Theorie; *or in the form of Driesch's theory;*[29] *or in the form of Oldekop's hierarchical monadism, where the parts and the whole rival each other.*[30] *"Totality" and its action have for him a phenomenological and not metaphysical character. Similarly, in a brief paragraph,*[31] *he repudiates* die sogenannte Zweckmässigkeit *and every teleological consideration. The positive power of finality lies in the "conservation of the whole," its optimal constancy, which is the organism's* Ziel *(goal) and not* Zweck *(purpose) . . . and which must not be taken in a realist or metaphysical sense but as a category of biological knowledge.*

K. Goldstein escapes the contradictions of organicism only by refusing to pose its problems. A few pages earlier, however,[32] *he cites and condones N. Bohr's and P. Jordan's reconciliation of microphysical indeterminism and the "acausality" of the individual organic being, whose actions always contain a personal and ungraspable factor. But from this reconciliation, Goldstein draws the argument for not posing the positive problem of causality and the mode of active dynamism of the "whole" in the organism. Before the living being that always strives to achieve an optimal state, one should not, we are told, try to understand how the behavior occurs and to explain it mechanically or teleologically.*

Organicism's "squared circle" appears clearly in the very title of E. C. Tolman's work: Purposive Behavior in Animals and Men. *For the word* behavior *signals that the author wants to affiliate himself with the antifinalist behaviorist theory, and the word* purposive *implies that, to be faithful to experience, the author had to admit finalism at least as an "as if." But he says in his preface that he detests the terms* purpose *and* cognition. *Tolman is an antifinalist, and he only makes use of these terms in a neutral and objective sense for which he refers to the glossary. Let us examine the glossary's definitions. A* purpose (fin) *is a "demand" for achieving or avoiding a certain type of goal-object; it can be objectively observed through certain modalities of behavior: persistence, docility, and so on. A "demand" is an "immanent determinant" of the organism, an innate or acquired urge to attain or avoid an object or a state, objectively definable by a certain type of behavior. Whereas Watson's behaviorism is "molecular" (behavior is a sum of muscular and glandular responses to stimuli), Tolman's behaviorism is "molar" (behavior concerns the organism as a totality). Behaviors, "though no doubt in complete one-to-one correspondence with the underlying molecular facts of physics and physiology, have, as 'molar' wholes, certain emergent properties of their own."*[33] *These new properties are, Tolman repeats, not only strictly correlated to physiological movements "and, if you will, dependent upon [them]," but for the description and in themselves, they are distinct from these movements.*[34]

Tolman's description of these "molar" properties closely resembles MacDougall's description of a typical finalist behavior: (a) an object or goal-situation is sought or avoided; (b) means-objects are employed; (c) the "shortest" means are preferably selected; (d) moreover, the behavior is persistent through trial and error and is docile, that is, teachable and perfectible. But whereas for MacDougall the finality of behavior implies a subjective, psychological reality behind observable appearances, for Tolman it is only a matter of an objective description. Even the apparent cognition *of goals and means has a purely "behavioral" sense; it designates the fact that the mode of behavior is a function of the nature of the environment. Tolman thus claims that his system is at once behaviorist (because behavior, although molar, is completely dependent*[35] *on stimuli) and finalist (because his descriptions of behavior, although objective and not "mentalist," are very close to those of MacDougall). Tolman does not negate subjectivity and consciousness* (qualia; raw feels), *but he admits that they cannot enter into a scientific construction, which seeks to coordinate, predict, and control; they belong to poetry and not to science.*

On the whole, Tolman's descriptions are excellent, and the value of his book is not marred because in some sentences here or there he expresses his conviction that finalist behaviors are dependent on physicophysiological phenomena. At most, he is forced to resort to burdensome circumlocutions (such as running back and forth behavior *to denote* awareness*)*[36] *and to the use of a glossary. It* *seems, then, that he proved the existence of movement by walking and justified the organicist attitude by his experimental output. Actually, he proved that in psychology as well as in biology, it is inevitable and indispensible to adopt* in fact *the finalist viewpoint, even if one assures the reader (out of prejudice or habit) that the finalist description is compatible with a mechanistic philosophy. Watson's philosophy led him both to questionable descriptions and to precious discoveries, because the false contains the true and the false. Tolman neutralizes Watson's philosophy by adopting* in fact *the finalist point of view. Because he sticks to descriptions, the philosophy he continues to proclaim has no significance or influence. But it is the philosophy that interests us here and, philosophically, it is clear that Tolman's point of view is untenable. Acts like "avoiding," "seeking," "choosing," notions such as "means," "end," "demand," "waiting,"* Sign-Gestalt-expectations,[37] *obviously imply an apprehension of sense and thus a consciousness. It is not difficult to do without the "word" consciousness, because the finalist notions used in the description already imply consciousness and the efficacy of consciousness. A* sign-Gestalt *that has a sense is thereby consciousness. By defining it through its motor effects, one does not reduce it to a pure mechanical cause, because the set of motor effects considered in a "molar" way is an* action, *not a set of movements, and the difference between an action and a sum of movements lies precisely in the conscious bond of action. We can experimentally observe that the animal does not operate before a* sign-Gestalt *like an automaton fitted with a photoelectric receiver.*

We have seen that bonds in general, always implicating a domain of survey and subjectivity, are never observable but only inferable and knowable. The misrecognition of this distinction is the key to organicist systems like Tolman's. Starting from the idea that science is pure observation, they have no difficulty showing that consciousness is never observable, because this is perfectly true and incontestable. But science is observation plus *knowledge, knowledge via observation. The observation of an automaton allows us to infer that there is no need to know it as autosubjective. The observation of an animal forces us to infer that it is conscious, to know it as a conscious being. So it is contradictory to say that its behavior is completely dependent, despite its "molar" aspect, on physical or physiological microprocesses. This would amount to saying that* it does not have *the mode of bonding of an aggregate or of a machine* and that it has *this mode.*

Merleau-Ponty critiqued both Tolman and Gestalttheorie, *and it is with some arbitrariness that we can classify his theory among the organicist doctrines. But he is very close to Goldstein, although his work is infused with deeper philosophical considerations. Like the organicists, he affiliates himself with the Kantian conception of finality*[38] *and of the phenomenon in general. Like the organicists, he leaves unresolved the problem of the dynamic relation between the overall behavior and the physicochemical processes in the organism and has a tendency to believe that the categories of interpretation or of description are as such categories of explanation of organic life in itself, as if living beings belonged to a universe of thought and not to a universe of realities.*

After very fair criticisms of pseudo-forms, that is, of Köhler's physical Gestalten *(the simple outcome of the balancing of parts), Merleau-Ponty defines his own concept of form. "Form cannot be defined in terms of reality but in terms of knowledge, not as a thing of the physical world but as a perceived whole."*[39]

This definition, it seems to us, is inexact. It applies to the conscious image *of a form but not to the form itself. If the form can only be defined as an object of perception, then we are doomed either to an infinite regress or to the oldest and least scientific idealism, disguised as neorealism. Merleau-Ponty does not draw the distinction between what is primary and what is secondary in psychological consciousness. We have seen the point at which it is essential to grasp that the perception of external beings, enabled by cerebral auxiliaries, does not form part of the primary texture of consciousness as subjectivity. As a domain of absolute survey, an organic form is altogether different from a physical* Gestalt, *and yet it is not a "perceived form." It is an abuse of language to say that the organic form is "perceived by itself," as though it had to present its own image to itself, like a man who looks at himself in a mirror instead of looking at others. It is an abuse of language to consider the autopossession of self, the "for-itself," the autosubjectivity of every being as self-knowledge or self-perception. This "texture-knowledge," this primary consciousness, is not knowledge; it is being. Perception's mise-en-scène must not be transported into the absolute survey of form-being and of activity-being. If it is true that a phenomenon is not an appearance, why continue to be duped by the etymology of the word* phenomenon *and assume that it implies presentation and perception?*

Let us imagine three humans A, B, and C on the model of Ripolin's famous poster. The first, A, is just an automaton but very sophisticated, made up of metallic cogs and dynamic systems of equilibrium. B is a living man but is deaf and blind and even temporarily deprived of every psychological life in the ordinary sense. The third, C, observes the first two. The first is certainly not a true form. Its "form" is constituted as a whole only in C's perception. It does not maintain its structure on its own, and it requires external maintenance and repairs. But B, an organism without psychological and sensory consciousness, is indeed a true form, because he is living and can be distinguished from a corpse, and because his organism actively maintains its structure (e.g., the stomach does not digest itself and the neural cells do not chemically degrade). This form does not depend on C's perceptive image of B. B's brain has a proper form and activity, which are no doubt less "molar" than if B were not temporarily unconscious, but less "molecular" than if he were dead. *Our three humans represent three levels: physical, vital, and psychologically conscious.* Gestalttheorie *as much as mechanism seeks the unity of the three levels by starting from A. Merleau-Ponty as well as idealists seek this unity by starting from C's interpretations. We seek it by starting from B, or from C as living, because B as a living organism is the type of normal and in fact universal being: it is an autosubjective form, an absolute domain, self-surveying, which is synonymous with "self-perceiving." A is merely a step-by-step assemblage of elementary beings. As to C, it is identical to B, with the difference that he perceives A and B through healthy sensory and cerebral assemblages. This perception is secondary relative to C's life: to perceive, to be psychologically conscious, one has to be alive. To have a conscious "image" of another being, one has in the first place to be a "true form." This perception is even more obviously "foreign" with respect to B, the perceived being, whose life is totally indifferent to the interpretations imposed on it. No one can seriously maintain, for instance, that it is C's observation of B that keeps B's stomach from digesting itself.*

The conception of epistemological, idealist, or neorealist monism, which turns perception and the perceived being into a numerically single being, is an insupportable paradox that, from Berkeley to American neorealists, has not ceased to muddle everything.[40] *It is all the more dangerous because it inextricably mixes the true and the false. It is perfectly true that C's psychological consciousness is not a kind of camera in which C's perception of B would be similar to a material image that copies B. Perception differs from a photograph in two ways: (1) as knowledge of B, of B's "sense," it transcends physical and chemical phenomena, which are based on visual sensation as "observation," and, through its intentionality, it is indeed at "one" with B, even if it is numerically distinct from him in its being; and (2) in its being, as C's state or activity of consciousness, it also differs from a photograph, because it forms part of the organic form that C is, a domain of equipotentiality and of absolute survey, which "lends" its subjectivity to perception. It is nevertheless absurd to identify perception and the perceived "other." When perception allows one to apprehend the "sense" of the other, there is no duality between the sense one grasps and the sense that is grasped, for sense is beyond space-time and belongs to the region where the numerical identity of similar beings reigns. Sense is beyond the subject–object categories. But perception, through the whole "cuisine" of sensation on which it rests, is nonetheless C's adventure, his act. It characterizes C; it is not the adventure, act, or state of B who is perceived. It cannot be used to resolve the problem of B's status as an autonomous living being.*

The significance of this analysis for the critique of organicism is clear. It is not because I consider the organism in one way or another, as a whole or a mosaic, that the problem of organic activity and its dynamic mode will be resolved. It is not the Erkenntnisgrund *that will afford me the* Seinsgrund, *nor will the perceived organism be the real organism. It is not because I, observer and "knower," have gone from an explanatory and physicochemical biology to a comprehensive biology that I can exempt myself from resolving the problem of dynamism peculiar to the organism and that I will be able to reconcile mechanism and vitalism or treat them as equal. The perspective of the understanding represents a first, indispensible step; but it is not everything. Von Uexküll's statement is perfectly on point: "Every organism is a melody that sings itself."*[41] *But Merleau-Ponty's commentary, that "this is not to say that it knows this melody and attempts to realize it; it is only to say that it is a whole which is significant for a consciousness which knows it, not a thing which rests in-itself," distorts the truth completely.*[42]

A melody is not a melody unless it is an "absolute survey" and not a mechanical juxtaposition of notes; we do not understand the melody, adds von Uexküll, by analyzing the ink with which the notes are imprinted, and the one who listens to the melody has to grasp it as a whole. Yet before the listener, there is the singer or the song that sings itself, that itself *dominates its own notes. A bird sings because it desires to sing, because it has a tendency to sing, in the same way that it had a tendency as an embryo to form its larynx. The bird's melody is the continuation of the "organic melody," that is, of the bird forming itself without witness or listener.*

If we refuse, on account of some academic purism, to turn signification into a force at the same time as an idea, we will never understand the real *organism and its* real *creative finality. "Doing biology" is not synonymous with "living." We realize that the current trend is to bring the theoretical biology of life and the life of theoretical biology closer together. It is true that to perceive as well as to sing a melody, one has to live it in some sense. So be it, but we should not exaggerate: to hear a song and to participate in a chorus are still two distinct operations.*

Let us move now to the positive part of this chapter. It is obvious that the Kantian and organicist conception of a finality by reflective judgment has to be renounced. Even if we do not retain from vitalism the idea of a specifically vital form, we should at least retain the idea of a force, a dynamic action on dominated and used physicochemical processes. It is a deeply ingrained prejudice to consider "unsophisticated" the thesis that turns the idea of consciousness into a force in the most precise sense of the term, a force truly capable of intervening in a physical process and diverting it. It took some courage for thinkers like Spearman, Heymans, and MacDougall to defend this thesis against the vast majority of philosophers. Nevertheless, this prejudice is justified less than ever today. Kant and Claude Bernard participated in a mechanistic, deterministic science, which represented the smallest part of matter on the model of celestial bodies subject to Newtonian action. It seemed just as incongruous to imagine that a vital or "psychological" idea could divert a molecule as to imagine Jehovah diverting the trajectory of a planet, which Newton did in his theological daydreams. This representation of the world is outmoded. The "particles of matter" are domains of action that become, in their interaction, a single domain and share their energy. The modern conception of bonds turns an interacting system into "a kind of organism in the unity of which the elementary constitutive unities are nearly absorbed" and which therefore acts as a systematic unity and not as a sum of elementary actions.[43] So, like the problem of the origin of life, the problem of the origin of so-called vital—it would be better termed "microorganic"—force no longer arises. Macroscopic organisms are progressively formed along the lineage of individuality of the universe, through colonization, dominated division, and hierarchical association of microorganisms, that is, of molecules. "Vital force" does not differ in nature from physical force, from the force of internal bonds of atomic physics's unitary domains of action, whose "force," as it appears in classical physics, is merely a statistical resultant.

The physicists who have insisted the most on the purely statistical nature of the *laws* of classical physics have also stressed that it is contradictory to apply to the individual what presupposes interactions (uncoordinated and simply added vectorially) among a multiplicity of individuals. And yet, even they have not always seen, or at least expressed clearly for the layperson, that force in classical physics, which also has a statistical and summative character, is as unlikely to apply to the dynamics of the individual as the statistical laws to the explanation of the structure of the atom. Macroscopic force (e.g., the attraction of the sun or of a large magnet) sums an enormous quantity of molecular actions; it is by its nature homogeneous to the individual force but its mode is very different. It appears as an unstructured quantity that can vary continuously. Because we are most often dealing with macroscopic physical forces, we are used to considering every force in this way, that is, as an amorphous, anonymous quantity with continuous variations; and when our thought bears on the individual force of a complicated organism, on the force of a tendency, of an instinct, of an embryological regulation, we do not recognize the fundamental identity of this force and the physical force, and we adopt either an ideal of reduction to macroscopic physical forces—which is contradictory—or the naive vitalist thesis of a specific vital force that differs in nature from the physical forces it controls. To resolve this problem, it is enough to regrasp the continuity of individual physical force and of individual "vital" force along the lines or fibers of individuality.

This truth could be perceived since the nineteenth century, and it partly was by some perspicacious thinkers like Cournot,[44] thanks to the very particular mode of chemical "force," that is, affinity. Cournot already drew a clear distinction between a macroscopic physics (of secondary laws) and a microscopic "infinitesimal, corpuscular or molecular" physics[45] of which dynamic crystallography and chemistry form part: "Whereas mechanical forces . . . engender effects that vary with distance according to the law of continuity, chemical actions only give place to abrupt associations or dissociations. . . . The chemical mass is measured by the capacity of saturation."[46]

We can say that chemistry was the first of "quantum" theories in which individualized forces became visible. The situation is much clearer today, especially since the wave theory of chemical bonds and of the capacity of saturation. The heteropolar or ionic chemical bonds (e.g., between Na^+ and Cl^-) can ultimately be interpreted through classical physics's continuous fields of forces, but that is not the case for homopolar bonds. How can two neutral atoms, for example, two hydrogen atoms, be united to form a molecule, and why is there saturation? "Even if in classical physics attraction forces between neutral particles were known, it would be quite impossible to understand why a third atom should not also be attracted by the two atoms already bound."[47] "But from the example of gravitation it can be seen how little the chemical forces with their saturation properties can have in common with classical forces."[48] Heitler's and London's theory links the binding energy to the exchange energy of two atoms, which is itself linked to the fact that the two hydrogen atoms have an antiparallel *spin*[49] and cannot be distinguished. Heitler suggests an imperfect way to represent this fact with classical or semiclassical images: he compares the two hydrogen atoms to two resonant vibration systems that exchange their direction of spin with some frequency of combination. But the exchange effect cannot be represented; it is a fundamental fact just like Pauli's fact of exclusion (two electrons occupying the same level must have an opposite spin).[50] Spin is in reality an indefinable state of the electron, a degree of intrinsic freedom, and it cannot be purely and simply assimilated to rotation or to any other structure or functioning of the electron.

So it is easy to see that the binding forces in a molecule are indissociable from a certain overall structure or, rather, because the term *structure* is improper, from a certain overall organization. A molecule is a whole in which the state of one part controls the state of another through an action that cannot be reduced to a causal, step-by-step influence, a whole in which the bonds are not absolutely localizable. The term *part* must not be taken in a strictly geometric sense. The exclusion formulated by Pauli's principle is not a "local" exclusion but a kind of indefinable *incompatibility*. For the physics of the individual, force is thus very different from a pure quantity. It does not have a sense simply as a vector has a sense (i.e., a direction); it retains a structure or, rather, a unitary organization of activities. Whereas macroscopic force can only "retain" a "Gestalt-form," the microscopic force is indissociable from a true form, from a veritable domain of survey. Nothing but habits of thought formed by statistical physics prevents us from imagining that a dynamism of the same type (a micro-macroscopic dynamism, to borrow P. Jordan's expression) can be indissociable from a much more complex form, from an organism in the ordinary sense of the term.

Merleau-Ponty spends a good deal subtlety to reject contemporary physics's royal gift to philosophy: "The fact that the physical system is imaginable today only with the help of biological or psychological models . . . does not accredit the chimera of a mentalist physics or a materialistic psychology."[51] "Mentalist physics" or "materialist psychology": the words are surely too strong. There is always something purely relational in the expositions (supported by equations) of "mentalist" physicists, more clearly than in MacDougall's psychological descriptions, reviewed and corrected by Tolman. But, all the same, it is quite interesting that today we are able to accurately grasp, through physics, the way in which effective and regulative "vital" or "psychic forces" are in continuity with molecular forces and can effectively guide them precisely because they have the same nature.

It must no doubt be understood that, in a complex organism, the specific forces that safeguard the unity of organization and of behavior, the formative instincts and the instincts tout court, do not act directly on the molar physical processes of subordinated systems. There would be an overwhelming disproportion between the energies put to work on both sides. Ordinary physical forces result from the addition of an enormous number of elementary components; in contrast, organic force is quantitatively on the same order of magnitude as the forces of molecular bonding. If organisms took millions of centuries to perfect themselves, it is precisely because they had to accumulate technical complications to dominate the molar and added forces indirectly, through hierarchical relays. It would certainly be puerile to imagine that it is the "organic force" that directly prevents the stomach from digesting itself or the living cells from fixing the dyes as easily as the dead cells. It would be puerile to imagine that the "force of enthusiasm" *directly* increases a man's productivity. It is quite probable that the biologist and the psychologist, in studying these kinds of phenomena, will always stumble on a physicochemical relay, on a "servo-mechanism" interposed between the order and the realization. But we should not conclude that from relay to relay, we go to infinity without ever discovering the point at which the "substituted chains" stop and allow the direct command to become visible. This point is probably situated beneath the cell's order of magnitude, at the level of the molecules used by cellular chemistry, as neomaterialists recognized. But whether it lies there or elsewhere, the moment necessarily comes when the command is direct. Expressions such as "Caesar forged a bridge" or "Khufu constructed a pyramid" are not, strictly speaking, figures of speech. They are condensed but literally exact expressions, and it would be more artificial to say that the will of Caesar or Khufu played no role in the movement of the laborers who alone, in the eyes of a superficial observer, constructed the bridge or the pyramids, just as in the eyes of mechanists or organicists it is the physicochemical forces alone that construct the organism.

To make matters clearer, let us resort to a myth.[52] Two inhabitants of Sirius, armed with telescopes of great yet limited power, observe the planet Earth and discuss the nature of beings that can be found there. The first, P, makes a decisive discovery; he notes that fires are lighted on Earth and are much more numerous in the cold and rainy regions than in the warm and dry regions. Because this phenomenon is contrary to the laws or the probabilities of physics, he concludes that there exist on Earth beings endowed with vital force who can struggle against these laws. But thanks to an improvement in his telescope, the second inhabitant of Sirius, S, discovers that to light these fires in winter, Earthlings make use of phosphor matches kept in their pockets and ignited by friction. He concludes that his colleague is mistaken and that, despite initial appearances, everything takes place on Earth in conformity with the laws of classical physics. P and S are both mistaken: P because he overlooks the existence of physicochemical *relays* in the ignition of fires, S because, discovering these relays, he extrapolates imprudently and fails to see that they are attached to an intelligent and sensible intention, which—if we do not wish to conflate everything in the endless chain of a universal determinism—should be understood as dynamic in and of itself.

The invention of matches or the intention to use a match at a given moment cannot be explained in the same way as the combustion of a match. The formula *natura non nisi parendo vincitur* cannot be absolutely true, because if all beings always "obey" (in the deterministic and nonaxiological sense of the word), it is not clear how nature could be vanquished. An intention has to be dynamic and, from a microswitch operated by a primary bond, has to effectively orient the deployment of macroscopic forces.

Spiritism (or the vitalism of the eighteenth century) is false, not because it attributes the character of a force to the mind or the vital direction, but because it endows it with the character of a macroscopic force that could act directly on phenomena at our human scale and directly realize an intention. The efficacy of consciousness cannot be denied, and epiphenomenalism is nothing more than an academic theory. But consciousness is effective only through the organic and extraorganic technology; spiritism is childish because it believes that consciousness is effective outside of every technology, not because it sees in consciousness a true force. Mohammed believed that the intensity of his faith could order the mountain to come to him. Faced with failure, he had the good sense to go to the mountain. He thus demonstrated that faith could accomplish an action when it undertakes it in the right way. A modern engineer who can add to organic technology the whole extraorganic technology of a protracted civilization can even transport the mountain or unite two oceans with a channel. It is indeed consciousness or, if you prefer, "spirit," "faith," that is the primary motor: a microspiritism is thus true. "A man may have faith as a mountain," Butler notes, "but he will not be able to say to a grain of mustard seed: 'Be thou removed, and be thou cast into the sea'—not at least with any effect upon the mustard seed—unless he goes the right way to work by putting the mustard seed into his pocket and taking the train to Brighton."[53]

The "magical" conception, like the "spiritist" conception of consciousness as a force, consists in believing that this force is not subject to the restriction of a certain order of magnitude or that it is not subject to the use of technical means when it tries to surpass, in its efficacy, a certain order of magnitude. But within the limit of this order, a part of the "magical" conception becomes true. In a domain of absolute survey, it can be said that there is magical participation of parts, action at a distance, omnipotence of thought, mnemic invocation similar to the invocation of "spirits," and immediate incarnation of significations. Because we are true and—in the sense in which contemporary physics employs this term—microscopic individuals, our immediate experience is naturally that of the magical mode of efficacy and force. By itself, objective and scientific experience teaches us about the relays that our action uses. But our direct experience suggests that it is our will that moves our arm, that it is indeed our intention that provokes our movements and our ideas. In the belief in magical action, the abuse has consisted only in extending to the outside world what is perfectly true of our domain of survey and direct inspection. The extension of extraorganic technology has made true the error of magic; it has given us power over the outside world, because it consisted precisely in complying with the very conditions that already made possible the passage from rudimentary organisms, like the molecules and the virus, to complex organisms. The "flying carpet" of *A Thousand and One Nights* is magical; a plane is not, even though it realizes the same dream and the same idea, and even though it is indirectly the distant effect of this idea, without which steel, aluminum, wood, and fabric would surely not have assembled themselves into a plane.

If one refuses to believe in the truth of "magical action" for the primary domains, one will sooner or later be forced to pay for it. One will succumb to the temptation of wrongly believing in this action to resolve questions in which it should not intervene. It is typical that several organicists pay for their purism in this way: "That the mind accords with the laws of nature, and that matter knows how to translate the mind's will . . . how can we not see at work here the technique of every mysterious operation and, it must be said, a magic? There is a magical power of life."[54] The refusal to believe in the dynamic character of primary consciousness necessarily entails the belief in parallelism, in preestablished harmony; and this belief entails in its turn the belief in a causality by magical participation, even in the macroscopic order. The same holds for the theories of perception that refuse to see the whole dimension of energetic interaction it contains. Lovejoy was able to demonstrate that the neorealist theory of perception, which refuses to admit the numerical duality of perception and the perceived being, entails a profound disruption in ordinary causality.[55] We showed[56] that Bergson's theory and thus the theory of critical organicists and phenomenologists who conflated Husserl's and Bergson's points of view entailed a magical conception of causality.[57] It is not worthwhile to deny the dynamic character of the elementary mental act only to then reach an out-of-place magical theory.

Because the macroscopic is merely an accumulation of "microscopics," the mechanical merely an accumulation of "organisms," there is a difference of mode and not of nature between the physical forces and the organic or conscious forces. The main difficulty that vitalists encountered and that perplexed the organicists—"How can one admit that a vital or psychic force without material support could intervene on physical forces from which it differs in nature, on physical forces that are inseparable from the material masses that carry them?"—this difficulty no longer exists because matter has been resolved into domains of action whose essential traits are identical to those of the domains of absolute survey. We can say that every force has a mental origin and that Leibniz was right (contra Köhler). The striking analogy between the modes of action of a force and those of a value do not prove, as Köhler believes, that value is reducible to force, because force is merely a macroscopic resultant, at least as he understands it.[58] It proves that every dynamic tension can ultimately be reduced to the action of an "ideal." More precisely, force (as the bond of the domain) manifests the "metaphysical transversal" that makes the true form of such a domain indissociable from an idea or a transspatial theme. This idea in its turn can aim either for a universal essence or for an essence that has already been transformed into a specific mnemic theme (instinct) or an individual one (acquired tendency). The act and the actualization that perfectly obey the idea or theme for which they aim would manifest its dynamism without the I-act of the "agent" experiencing it as an "impression of force." But it is enough—this is practically always the case, owing to the internal hierarchical structure of beings—that a gene be actualized for the force to be experienced as well as manifested, for the manifestation of an idea, an instinct, a haunting memory to be impeded by an external or internal obstacle and the impression of force to appear immediately in both the impeded being and the impeding being. Then there is struggle, the effort of two beings or two subindividualities in conflict, whose resolution will be the constitution of a more unitary system. This is why, if force in its essence results from the physicochemical nature of the unitary domain, the impression of force results from the relative alterity of two interacting domains or of two subindividualities in a complex domain. The felt force is always the "ideal" or the "virtue" of an "other" experienced from the outside. When the alterity is absolute, the "other" seeks to eliminate me. When it is relative, the other acts by trying to "convert" or "persuade" me,[59] and I act on him in the same way. We think we are speaking in metaphors when we apply these psychological descriptions to force in general, when we speak of the "force" of an authority that persuades us and converts us to its ideal. But in fact we discover here the truly primary nature of force. It is the force of statistical physics that is, if not metaphorical, at least "degenerate." The pressure of a gas, of a liquid, or of a Gestalt-form out of its state of equilibrium is the outcome of billions of elementary actions, each of which manifests the primary force generated by the obedience to an ideal norm. Just as the elephant is, despite appearances, more "microscopic" than a soap bubble, so the force of instinct—when it is impeded in its development—or the force of a defied ideal is more primary and more "elementary" than the force of an overinflated balloon that bursts.

19

Psycho-Lamarckism

Because we are not studying the theories of evolution but the theories of finality, we will only consider the "psycho-" form of Lamarckism. We can ask as a matter of fact whether every Lamarckism is not psychological.[1] By psycho-Lamarckism, we mean the conception that explains the internal finalist assemblage and the de facto adaptation of organisms to their environment or to their living conditions as the result of an accumulation of direct individual efforts, psychological in nature and similar to the conscious effort.

In general, the notion of adaptation does not refer to a particular doctrine. As G. G. Simpson notes, "it is a truism that all the organisms can live under the conditions under which they do live and that they could not live under other sets of conditions that exist. To this degree, at least, and without any teleological implications, adaptation is universal."[2] By the same token, the notion of preadaptation can be interpreted in a mechanistic sense or in a finalist sense. For psycho-Lamarckism, adaptation is teleological. More exactly, (1) it is first of all the realization of an end by the individual and (2) it implicates an integration of individual efforts through memory and habit, which can become overindividual and pass from one individual to another. This is roughly the thesis of E. Hering, Samuel Butler, Cope, Pauly, MacDougall, Vignon, Pierre Jean, and so forth. Even though he essentially adopts a psychological conception of life and speaks of the "current of consciousness launched into matter,"[3] and even though he endorses neo-Lamarckians' recourse to a psychological cause to explain evolution, Bergson cannot be ranked among psycho-Lamarckians because he critiques—rightly, as we will see—the idea of an accumulation of individual efforts as a final cause. "The truth is, it is necessary to dig beneath the effort and look for a deeper cause."[4] The thesis he espouses is at bottom a combination of Eimer's theory (invoking orthogenesis) and of psycho-Lamarckism. He retains from Eimer's theory the idea of an internal principle of direction in life (overindividual and even overspecific), rejecting the physicochemist interpretation of orthogenesis that Eimer seems to adopt; he retains from Lamarckism the notion of the psychological principle of development, rejecting the idea that this psychological principle is the implementation of individual efforts.

To be sure, we cannot rank authors like Richard Semon, Rignano, and Bleuler among psycho-Lamarckians as we sometimes do.[5] They have something in common with psycho-Lamarckians: they draw a parallel between psychological problems and biological problems, and they see especially in memory the key to heredity and development as well as to individual psychology. But we surprisingly discover that, while speaking of the "psyche" and of "psychoid," they reduce psychological memory and activity to pure energetic and physicochemical phenomena, a reduction that makes them slip back into all the contradictions of organicism.

In contrast, neovitalists like Driesch often evolve in such a way that they come very close to the psycho-Lamarckian viewpoint. In principle, by insisting on the specificity of the vital fact (force or entelechy), vitalism refuses to identify it with psychism: "Life is an original and irreducible reality."[6] Entelechy, says Driesch, does not have a mechanical or physicochemical nature, but the opposite of "mechanical" is simply "nonmechanical," which is not "psychic." On the other hand, Driesch, like Bergson, shows that the accumulation of small individual efforts, apart from postulating the doubtful heredity of acquired traits, could at most account for adaptive details in a given type but not for the type itself and its formation. From the histological perspective, he notes (following C. E. Baer) that the ringed seal can be as adapted as a higher vertebrate; it is nevertheless a lower type.[7] But he recognizes that Lamarckism, especially as A. Pauly formulated it, has essentially a vitalist inspiration and, on the other hand, in his last expositions, he did not postulate an entelechy in the higher organism but rather a hierarchy of entelechies crowned by the conscious "I" or the "objectal psychoid," "which uses the brain as a keyboard."

Strict vitalism, which is forced to supplement the immaterial factor of life with an immaterial factor of psychism, violates the law of the economy of hypotheses; the borders between the finalist regulations of the organism and the finalist regulations of conscious behavior are so clearly variable that vitalism will always be tempted to slide toward animism or psycho-Lamarckism.

Psycho-Lamarckism contains important elements of truth: it recognizes the finalist character of organization, instinctive behavior and conscious behavior, as well as the fundamental unity of the three levels; it recognizes that finality is inseparable from the mode of being of consciousness or the psyche in general. Invoking habit and memory as superindividual phenomena, it is logically forced—although it does not always realize this logical consequence—to conceive habit and memory as transspatial and as irreducible to simple material traces, thus as a kind of thematic potential capable of dynamically guiding a structuration. Psycho-Lamarckism may be right or wrong to admit the heredity of acquired traits and the integration of individual habits in the potential of the species; in either case, it cannot use spatial models—which are obviously inconceivable in these circumstances—to understand vital activity. This is so true that the famous Russian biologist Lyssenko, while proclaiming himself the champion of materialism, had to follow the internal logic of Lamarckism—which he adopted to the extent that he critiqued genetics and neo-Darwinism—to the end and to make patently finalist declarations.[8]

Nevertheless, psycho-Lamarckism is not as such an acceptable doctrine today, either for biological science or for a philosophy of finality. For its scientific critique, we refer to the valid objections of neo-Darwinians.[9] The most persuasive among these—beyond the classical objections drawn from the scientific nonexistence of experiments on the heredity of acquired traits; from the impossibility of a hereditary transmission of the work instinct among the sterile Hymenoptera workers; from the internal contradiction of a theory that postulates that a species is sufficiently plastic to undergo an action and sufficiently stable to retain it during an immense duration—are (1) the fact that mammals with perfect internal regulations, whose germinal cells are protected against every variation in the environment, had to evolve more slowly than the other living beings, which is not confirmed; (2) the fact that a multitude of traits cannot be due to individual usage (teeth are primed from the embryonic phase when they are not used, and individual usage can only deploy them mechanically; organic camouflage clearly does not derive from any individual usage or effort); and (3) nonadaptive orthogeneses, more difficult to explain for Lamarckism than for neo-Darwinism, which can at least invoke a genetic connection between an unfavorable factor and a favorable factor.

The philosophical weaknesses of the doctrine are of particular interest to us. There is at least a propensity, especially among Lamarckians with literary inclinations, to slide into magical and spiritist conceptions of finalist action. Romantic German philosophy and psychology, inspired by Lamarck via Goethe and Schelling, are very typical. For Carus, for example, feeling as "psychic force" is capable of directly controlling the organic functions and directly modeling facial features.[10] Likewise, in a utopian fantasy,[11] Bernard Shaw thinks of himself as a Lamarckian and the disciple of Samuel Butler when he attributes to "faith" and "will," identified with the *life force,*[12] the most extravagant direct action on the organism: prolongation of the duration of life and suppression of the organs useless for thought.

On the other hand, psycho-Lamarckism falls into a serious error with respect to the connections of finality in internal circuit and finality in external circuit, of organic finality and psychological finality in the ordinary sense of the term. As we have seen, finalist behavior in external circuit, which ordinarily presupposes the deployment of the nervous system and often the use of tools, is an extension of organoformative finalist activity: the act of searching for sugar when one is hungry extends in the external environment the organic act of storage and release of sugar in the internal environment. Before the evident resemblance of the two acts, it is perfectly legitimate to start from the act in external circuit and its patently finalist character to ascend *in thought* to the organic act and to reach a conclusion about its equally finalist character. But we should not to confuse the direction of the progress of *philosophical reasoning* with the direction of the progress of *the real formation.* Habit extends instinct by adapting it to the thousand circumstances of the environment, and it presupposes instinct; psychological invention prolongs and presupposes organic invention. We can draw conclusions about the *nature* of instinct or primary organic invention by starting from the nature of its extension; we can conclude that the primary instinct has to be essentially finalist and autosubjective, like habit or psychological invention. But we cannot draw conclusions about the *origin* of primary instinct. It is even contradictory to make what is prolonged emerge historically from what prolongs. This is nevertheless the mistake that neo-Lamarckians commit. Samuel Butler starts from the art of the pianist, which has become unconscious by dint of practice: "Is there anything in digestion, or in the oxygenisation of the blood, different in kind to the rapid, unconscious action of a man playing a difficult piece of music on the piano?"[13] So, he concludes (passing from the problem of nature to the problem of origin), it is impossible to believe that these operations "should be made without endeavor, failure, perseverance, intelligent contrivance, experience, and practice."[14] "Who," then, practiced in them? The continuous individual formed by the succession of millions of individuals, superficially distinguished by the minor incident of fertilization or birth.

It is first of all necessary to recognize the profound truth of Butler's thesis. It retains its whole value so long as only a problem of nature is at issue. It hits a snag with the problem of origin and the problem of the "who." The subject, the agent, the practitioner of the external circuit, is the conscious "I." Can it, by its efforts, create its own support, in other words, the *x* of the organic individuality and even the *x* of the species that dominates this organic individuality? Can the habit of sucking create the instinct of sucking, the instinct of swallowing, of digesting and of forming a stomach and a digestive tract for oneself? Can the habit of making provisions create the instinct of hoarding, then the formative instinct of organic reserves of sugar or fat? Can the sexual habits of a male individual create the instincts, then the sexual organs, of the male? And "whose" habit is it that harmonizes the instincts and the organs of the male and female? It is indeed clear that psycho-Lamarckism inverts the real order. If our tools are akin to external organs, and vice versa, if our organs are akin to tools (for the problems of nature, the order of comparison matters little), it is in fact the tools that presuppose the *existence* of organs and not the reverse (for the problems of origin, the order by contrast matters a great deal).

Neo-Lamarckians were deceived by the phenomenon of passage from the conscious to the unconscious, which seems to bring habit close to instinct. But active habits, at least within the bounds of our experience, become "unconscious" only in the manner of a psychological "other-I." They remain in the domain of the psychological in the ordinary sense of the term. A habit never assumes the character of an instinct nor especially the character of an instinct that forms organs; it never enters into the region of biological autosubjectivity. Secondary consciousness is never transformed into primary consciousness.

This error led neo-Lamarckians to another that we have already critiqued, the "shameful pan-psychism" that represents the primary consciousness of plants and of organisms without a nervous system as a diminished, evanescent, and vague psychological consciousness. The terms that underscore the resemblance between organic consciousness and psychological consciousness, like the term *psychoid,* inevitably turn into "diminutives" in the minds of those who use them.

Habit, learning,[15] cannot be the primary element of organic finality nor, moreover, of finality in general. Habit is an auxiliary of finality, an accessory channeling, an accommodation of subordinated details. Isolated from a principle of higher finality, habit always risks losing the general point of view to confine itself to a small domain of accommodation, often by creating adverse step-by-step "adhesions." If our living cells were to become completely "habituated" to their immediate surroundings, organic finality would quickly suffer: postoperative adhesions are one example of this. Let us admit that in a limited number of cases, the total organism (or the *x* that lies behind the organism) leaves the care of adjustments to cellular habits (e.g., the orientation of bone trabeculae, the details of capillary or venous anastomoses, the innumerable small adaptations that endow a plant with a specific form much less stereotyped than the form of an animal); it can generally do so without endangering the whole. An organic "interventionism" prevents, for example, the muscles of the uterus from atrophying even though they do not work and the muscles of the heart from swelling even though they work ceaselessly. One psycho-Lamarckian writes, "The cells that formed the knee did not invent it in one stroke, logically, methodically. . . . They empirically discovered it, through innumerable trials and errors, with small successive advances, like the men who discovered the smithy and agriculture."[16] This is perhaps true for some details of the articulation of the knee, but the general schema of an articulation must undoubtedly precede the minor tunings. The cells of the knee surely could not invent the entire bone system and muscular system. A sum of minor tunings does not amount to an invention. Psycho-Lamarckism often has a tendency to become a kind of monadic or "communal" theory of organisms. For psycho-Lamarckians, habit and adaptation reign in the relations between cells as well as in the relations between organisms or between organisms and the environment: "Given the multiple potency of cells [the cells of the embryo], the competition of cells, the symbiotic relationship of cells, and the tendency of cells to differentiate in accordance with functional demands, we may conceive how orderly development and regeneration may issue as a resultant of these component factors."[17] But these factors would no doubt produce an *arbitrary* organization,[18] just as assembled humans always end up producing more or less ordered institutions, but not an organization *that is consistent with a specific, well-defined type*. The organism looks as little as possible like a democratic and liberal society. It bears witness to an initial planning. Pure cellular adaptations, even helped by mnemic associations that integrate "the small successive advances," cannot explain organic finality, any more than associations of ideas or associations of reflexes can explain the finality of behavior.

It is true that psycho-Lamarckians also introduce "need," without ever delimiting precisely the respective roles of "need" and learning. It is on account of "need" that the muscles of the heart and those of the uterus retain their form despite the difference in exercise.[19] But what is this "need" that is invoked as a principle of explanation and that does not act via the use or nonuse it determines, as in primitive Lamarckism, but *despite* this use? It can only represent a kind of magical efficacy or the presence of a plan that is completely transcendent to individual efforts.

The word *need* has a double sense. It can signify (1) ideal demands of a being or a system—thus a combustion engine "needs" gasoline and a carburetor to function—or (2) the psychological state of tension or drive[20] of a living being that lacks something, for example, water or sugar. If psycho-Lamarckism invoked need in the *first* sense, it would no longer be a genuine Lamarckism; its finalism would be transcendent and no longer psychological. Need is the reason and not the cause of the organic structure, and it is effective *despite* the causes (e.g., *despite* its incessant beats, the heart does not swell up like an athlete's bicep). Lamarckism can legitimately invoke need only in the *second* sense. From this perspective, need is plainly a secondary phenomenon relative to the organic type, secondary like the minor adaptations of learning. It is a phenomenon that corresponds to the passage from the activity of one subindividuality to another or from one semi-independent area to another within an organism. Need has to retain the general structure of activity. The need to drink is a "message" transmitted from tissues to the central nervous system through complicated physiological mechanisms (they were studied by Cannon, Montgomery, Bellows, and Richter) in which hypophysial actions intervene. Psychophysiological need is therefore an accessory and useful assembly carried out by the organism; it presupposes a finalist plan, it cannot be used to explain this plan. The sophisticated engines regulate their own supply; the auxiliary mechanisms that ensure this automatic adjustment play in sum the same role that psychoorganic needs play. This is an improvement that cannot be used to explain the invention of engines in general. Thus the numerous theories that believe they explain finality by interpreting it as "the causality of need" argue in a vicious circle.[21] The drive always presupposes need. The drive acts in part as a cause, in an auxiliary chain established by the organism and that in effect resembles the chain of servomotors in mechanical automatism. The resemblance of automatic regulation and of regulation by need is, moreover, very imperfect. Although the drive is a simple auxiliary in the organism, like the mechanisms of regulation in the machine, it rests on a more fundamental property of organisms: the ability to invent according to need. The drive is not a pure cause a tergo; it always comprises an element of "searching" similar to the search according to a norm. But this confirms our preceding conclusions: an initial "planning" is indispensable for understanding the experience and action of psychoorganic need.

We are at any rate outside psycho-Lamarckian theses. The case of need parallels that of memory and habit. Memory is in general a fundamental phenomenon, but *learning*[22] memory is one of its derivative modes. Similarly, need is a fundamental reason for organic structures, but the drive is a derived psychic mode. Psycho-Lamarckism can only be maintained as a modest complement to a biological Platonism. Needs depend on the organic type: an insect does not have the same needs as a mollusk, an herbivore as a carnivore. And it is impossible to maintain that the specific needs and the types themselves are due only to an accumulation of individual habits. The finality of the species and especially of the type is not a sum of individual finalities. This is obviously the case for reproduction: it is in itself onerous for the individual. What individual could the advantage of reproduction be an experience of? Of progenitors? Plants and most animals derive no joy from family life. Of the progeny? For it, reproduction was beneficial; it allowed it to be born. But birth is not exactly an experience, and at any rate not an experience of reproduction. The need and pleasure of reproduction can only be a superadded "means" in the service of the primary finality of the species. And it is naive to extend need (in the form of an unconscious Will to reproduction) to the species itself, and especially to Life in general, when need is meaningful only in the "passage" from one individual to another at the interior of the species.

Pre-Darwinian biology of the nineteenth century distinguished between the "Aristotelian idea of the harmony of functions and the coordination of all parts of the organism in view of the functions to be fulfilled" and "the Platonic idea of the type of organization."[23] Some measure of truth subsists in this distinction. It is precisely all the elements of the "type" that the Lamarckian principles cannot understand. When they almost unanimously abandon Lamarckism, modern biologists are of course very far from thinking about Platonist types. And yet it is remarkable that for fifty years, biology, most often in the guise of physicochemical interpretations, has come closer to a Platonist finalism than to an Aristotelian or Lamarckian finalism. Genetics and mutationnism imply the permanence of "types" that are not essentially adaptive. They admit the passage from one type to another, independently of every drive and every individual effort. The theory of preadaptation goes in the same direction, because, through preadaptations to a certain environment, need is satisfied independently of every drive and of every individual effort. The possible relation with an environment precedes the real relations; the "prospective functions" precede the "realized functions."[24] The organ precedes the function, contrary to the Lamarckian axiom that "the function creates the organ."

As Simpson has ingeniously demonstrated, phylogenetic evolution can be represented as the encounter of a certain number of "types" with a "grid" of possible adaptations in space and time. Simpson maintains that, in contrast to specific differentiation, "rectilinearity" in evolution, so-called orthogenesis, or the phyletic evolution in general (e.g., the evolution of the stock of equines) can be explained by the fact that a more or less linear (or *pathlike*)[25] evolutionary outline is imposed on the linear structure of the "grid" of adaptations. Phyletic evolution is already more fundamental than the ever-minor specific differentiations, and it cannot constitute their sum. But more fundamental than the phyletic evolution is the evolution that Simpson proposes (significantly) to call "quantum": a "relatively rapid shift of a biotic population in disequilibrium to an equilibrium distinctly unlike an ancestral condition."[26] "[This mode constitutes] the dominant and most essential process in the origin of taxonomic units of relatively high rank."[27] It introduces pronounced or radial alterations in the physiological system; it implicates a *nonadaptive,* then *preadaptive* phase—normally by fixing mutations in a small population—before the new adaptation to the new zone of equilibrium.

Simpson's synthesis shows that, regardless of the interpretations, and even if we do not adopt the author's neo-Darwinian views, psycho-Lamarckian adaptation and finality through need, effort, and learning play an extremely reduced role in evolution. In Simpson's mind, this synthesis eliminates finality in general, because for him the type is produced by fortuitous mutations, on one hand, and by the selective effect of the grid of adaptations, on the other. Simpson, as we have seen, does not believe in orthogenesis as a distinct and intrinsically guiding factor. But we do not find in the facts he introduces or discusses any reason to follow him on this point; quite the contrary. It can be said without paradox that this neo-Darwinian, like countless supporters of the theory, believes too much in *utilitarian* adaptation in the guise of selection. The very complicated and often extremely refined ornamental organs cannot be explained by a psycho-Lamarckian adaptation or by an orthoselection along a path determined by a grid of adaptation between two zones that are impossible for the species. How are the brilliant and complicated wings of the birds of paradise, morpho butterflies, or uranias adaptive? It is just as inconceivable that they were produced by selection as by "accumulated individual efforts"; they are not conceivable in any of the three evolutionary modes that Simpson distinguishes. The weak physiological significance of their differences would lead us to place them under the rubric of "specific differentiation." But the complication of their outlines prohibits us from contemplating an origin, whether by minor adaptation or by mutations and segregations due to chance: an enormous number of orthogenetic mutations (hence an immense duration) would be required. They are "typical," and this is truly all that can be said about them on the plane of positive knowledge. Between two butterflies of closely related species and with practically identical ecologies, yet bearing very different ornamental patterns, there can be no "quantum leap," as Simpson understands it, like the one that transports the equines from the "leaf-eater" zone of adaptation to the "herbivore" zone. Between two decorative motifs, there is no law of "all or nothing."

Similarly, hypertelic orthogeneses are an objection against both Lamarckian adaptation and neo-Darwinian adaptation (*Gryphaea* valves, which in the end could no longer open; horns of the last Brontotheriidae; tusks of the last mammoths; backbones of some Permian reptiles; oversized antlers of megaceros).[28] This hypertelism is not an atelism: thematically, the hypertelic organs are always organs and not arbitrary aggregates. It is only an objection against a utilitarian and individualist finalism. It is compatible with an aesthetic and cosmic finalism that aims for the realization of the most varied types. It seems to manifest an "evolution-program"—according to Bulman's expression—that goes beyond the advantage of individuals.

The critique of psycho-Lamarckism (especially if the loophole of natural selection is rejected) leads us very closely to metaphysics, to the metaphysical element of reality. The type cannot be explained by psycho-Lamarckian action; and because it also cannot be explained by the formula mutations + adaptive selection, it remains only to accept it as a primary fact. The psycho-Lamarckian finalism of effort and of individual learning had the advantage of keeping transcendent metaphysics at bay. Samuel Butler recognizes the necessity of ultimately relating the individual and even the species to Life in general, considered as a single great Being, and in a curious book, he even relates Life as a whole, the known God, to an unknown God, the simultaneous origin of life and of the mineral world.[29] But this unknown God is distant, and it does not figure in *Life and Habit*. Pierre Jean believes that his psycho-Lamarckian theses allow him not to choose between "God and physics." In contrast, the finality of the "type" forces us to admit straight away a kind of metaphysical and theological initial emplacement, a primary plan(e). The historical character of the evolution of types and species must not veil their ideal and systemic characters. The types and species invent themselves in time, but this invention is guided, predestined. In some sense, organic memory is a ready-made pseudo-memory. Instinct, which has all the traits of memory, is no doubt a pseudo-memory *for the individual.* The pace of evolution compels us to go further and to admit that it is even a pseudo-memory *for the species.* It is a "reminiscence" memory, through the apperception of a "type," a memory with a determined program, a memory that is inseparable from a predestined invention. The small adaptations of the species are to the formation of types what the individual tuning of mimetism (through the nervous or hormonal system that acts on the chromatophores) is to structural mimetism.

Last, the recent evolution of the physics and chemistry of the individual makes it possible both to introduce a decisive argument against psycho-Lamarckism and to interpret the failure of psycho-Lamarckism in the sense that we have sketched out. Because no absolute barriers exist between the large molecules and the most elementary organisms, between chemical individuals and living individuals, the theories of organic evolution and the conceptions of finality must be equally able (at least by and large) to apply to "molecule-organisms," situated at the borders of chemistry and biology, and it must even be possible to extend them to the individuals of microphysics. One of the advantages of mutationism or of neo-Darwinism is that it seems to relate easily—as E. Schrödinger's theory shows—to the recent discoveries of contemporary physics. By contrast, it is not clear how psycho-Lamarckism could interpret behavior and the evolution of protein-viruses, because its key notions are all borrowed from the domain of ordinary psychology of man and animal: need, effort, learning, and so on. The crystallizable viruses and even the molecules and atoms can be interpreted as domains of activity that are "typified" by transspatial norms and by "prescribed possibilities," which subordinate a systematic framework and a systematic plan to their activity. The idea of a "typical" and Platonic finality is therefore very appropriate for all "organisms," in the broad sense in which Whitehead uses this word, whereas Lamarckian finality is only suitable to one particular category of higher organisms. And furthermore, this idea conserves the measure of truth contained in neomaterialism and neo-Darwinism: the mutations have a "quantum" aspect (this word is taken either in Schrödinger's sense or in the broad sense in which Simpson uses it), not because they are pure material accidents according to a blind causality, but rather because they are produced according to a systematic framework that comprises a discontinuous series of stable states. The living species do not form a table as strictly systematic as the chemical species in Mendeleieff's table, because activity in the higher organisms is less closely subject to norms than the activity of physical individuals and is complicated by all the secondary procedures that allow these organisms to adapt to diverse geographical and ecological environments; in contrast, microorganisms (atoms and molecules) have no need for such procedure, because they compose the physical environment and do not need to adapt to it. But the general system of types remains visible despite the innumerable variations on the great themes of organization.

20

Theology of Finality

The world is not without God, but God is not without the World.

H. L. MÉLVILLE, ***Vers une philosophie de l'esprit***

The situation of contemporary sciences forces us to turn from the problem of finality in the world to the problem of finality of the world. Theories like psycho-Lamarckism, which admit only an individual finality, are as insufficient as the theories that deny every finality. Finalist activities are *systematized.* Every unitary domain of action implicates a "metaphysical transversal," but the multiple "transversals" of various domains cannot be considered in isolation; they pose the same general problem. Let us compare a few typical structures of unitary domains. We obtain a table of this kind:

[I] strive to establish propositions consistent with [truth].
[I] strive to recall a [mnemic theme].
[x] (an embryo) actively organizes itself according to its [specific type].
[x] (a living species) actively evolves toward a [harmonic type].
[x] (a molecule) actively maintains its [typical form].

The isomorphism between these different cases is hardly questionable. There is always [an agent] that strives to realize [an ideal]. We have been able to observe on the basis of contemporary science that all the beings in the universe are domains of activity, of finalist activity, which take this general form. Only the "aggregates," the "crowds," are an exception and degrade finalist activity into a pure evolution toward an extremal equilibrium.

[An aggregate of *x*'s] passively evolves toward a [Gestalt or maximal entropy].

By its nature, science grasps in full only this final, degraded case. As concerns all the others, it studies the various modes of work and activity in space and time and ignores or tries to systematically ignore all that we have bracketed: the Agents as such and the Ideals as such. It tries to ignore the internal bonds and the sense of the activities it observes. Nevertheless, it directly informs us about the universality of finalist action, about the variety of its modes and their reciprocal implication. It thus indirectly informs us about the Agents and the Ideals, because it cannot in fact separate physics from the metaphysical "transversal," for the domains of activity are tied and unified only by their metaphysical component. Although it may not grasp the bonds, science is forced to take them into account.

The task of metaphysics is twofold: (1) it transforms scientific observations into a knowledge of bonds and senses (but this metaphysics is created, in part instinctively and most often implicitly, by scientists themselves, who cannot help being "realists") and (2) as metaphysics proper, it studies the general status of what is bracketed and the relation between Agents and Ideals or the Agent and the Ideal, because nothing allows us to affirm a priori the fundamental plurality of either.

This twofold task should not be performed too early, prior to scientific information and with the pretension of orienting this information. Countless scientific errors occur because scientists resort too hastily to an implicit metaphysics (according to task 1) and fall into a bad realism. Countless metaphysical errors occur because philosophy pursues the second part of the task too hastily and extends the errors of science's implicit metaphysics. We can and must establish, on the basis of the examination of scientific results, that sense and finality are everywhere, before turning to the metaphysical problems of the "I" or "God," that is, Sense or Logos. The Cartesian system, as is well known, inverts the order. It begins with a metaphysics considered as preliminary to a physics. It fails to recognize the axiological character of the "cogito." It fashions an ontology of the thinking Substance, on one hand, and of the Perfect or God, on the other, an ontology of the two expressions it should have bracketed because they do not represent immediate givens in the same way as the "work of thought." Undoubtedly the "work of thought" is in an immediate way not only here-now but here-now-I. But the "I" of this triple expression is not the ontological and substantial "I" in which Descartes thought he found himself straight away; it designates the Agent or the Acting.

With more precise insights and after three centuries of immense scientific progress, we can turn to the metaphysical problem (task 2) without entertaining many illusions. Every metaphysics in the sense 2, like every theology, is mythical. It is always necessarily a "work of thought" that takes place in the actual, in what is not bracketed, while claiming to place itself outside the total system it tries to define. Doing metaphysics, whether dogmatic or critical, always consists in pretending to be God or the Witness of God, in other words, the absolute Totality that deliberates with itself and takes us into its confidence. In the *Book of the Secrets of Enoch,* a Jewish apocalypse from the Christian era, the patriarch is lifted by the angels and sees in the seventh heaven God himself, who reveals to him the mystery of Creation and deigns to show him in detail how he acted on each of the six days.[1] At bottom, every metaphysician employs, without admitting it, the simple procedure of the ancient Jewish author.

This fiction is so implausible that the metaphysician should perhaps abandon it and occupy himself with other exercises. Nevertheless, because every mysticism has believed in the identity of the "I" and the absolute; because pantheism maintains a similar thesis; because Kantian criticism contains the germ of this idea as its successors and heirs have amply demonstrated; because idealism, rationalism, in short, the most varied metaphysics postulate it; because, on the other hand, according to scientific facts themselves, we do not know exactly what lies between the brackets on the left nor by consequence what lies behind the pronominal "I" that speaks and deliberates, we cannot rule out a priori that the fiction contains a certain measure of truth and that, to this extent, metaphysics is possible. It is quite curious that a physicist like Schrödinger ended up considering his "I" as "Atman." He did so in a highly questionable way and to rule out a contradiction in which he trapped himself, but his example at least justifies analogous fictions. It is already something not to be "naive" and to candidly acknowledge at the start that these two propositions, "I believe in the possibility of a transcendent metaphysics" and "I believe that at bottom I am identical to God himself," are indissociable and can only be justified together and exactly to the same extent.

So let us adopt the fiction openly by admitting at the outset the secret postulate of every metaphysics, which claims to discover it as a conclusion. We are the Absolute, we are outside and beyond the Whole of reality; we discern the secrets of nature and the formation of a world of real beings. We see how everything converges toward the success of creation. This is then what We think or clearly see in the divine Thought that is in fact at one with ours. We strive to create real beings. "Real being" implies "free being"; otherwise, only a unique and compact block would exist where nothing could be distinguished. "Free being" implies "free activity," and "free activity" implies the two terms agent and ideal. The ideal is furnished by the divine understanding that will allow the agent to "discern" the ideal it has to realize. We conceive all of this, which analysis has meticulously extracted, in a flash, as God does. The divine task has only begun, but from the start everything prepares it for success. A pure society of free activities-beings, in which everyone does the same thing by aiming for the same ideal, would lack charm and variety. A universe composed of electrons or a universe composed of pure angelic minds would provide the image of this society. This universe would not be truly a Cosmos; the freedom of beings would be exercised only in the contemplation of the Norm, not in the effort to rearrange a natural reality. Common space-time would not exist; there would be nothing more than a coexistence of "proper times" inherent to each activity. Yet, by their very nature as "activities," beings can colonize one another, because they are not impenetrable substances. Atoms and molecules are formed and realize a varied system of forms that has its own mineral beauty. Furthermore, a crucial bifurcation takes place in the same stroke: there are two kinds of laws, (1) the laws that fix the form of various molecules according to the possibilities "contemplated" by molecular individuals and (2) the laws that govern the superficial interaction of molecules among themselves. Accordingly, not only is a nature or a system of beings formed in this case but also a Cosmos, a World, in the geographic sense of the term, with the fortuitous and the accidental, which can act as support and habitation for more sophisticated beings-activities. Life cannot be created distinctly, because all beings are already living and conscious in the fundamental sense of these terms: all beings are forms that actively maintain themselves. For organisms proper to emerge, it suffices for colonization to gain ground and for another bifurcation to appear, akin to the first and already virtually enclosed within it: that which separates the proper nature of each organism and its cosmological actualization or, in other terms, its typical memory and its realization "here and now." Afterward, the hereditary species and the reproduction of individuals emerge. Memories are from then on interposed between the sought-after Ideals and the Agents; they accentuate their specific character and, later on, their individual character. The nervous system, an organ that at first allowed the minor adaptations of organisms to the cosmic, ever-changing environment, soon becomes the occasion of a kind of total reflection of all creation through the perception it enables, then through the constitution of a universe of symbols. Perceptive consciousness is not, moreover, a novelty; it is a simple adjustment of the primary subjectivity of beings to fulfill a particular function. Social life and the pooling of conscious individual activities enable the constitution of a supraindividual memory, which is not specific memory and which further heightens the autonomy of beings at the same time as their power of realization. Nevertheless, this power of realization does not become dangerous for the beings that possess it because they remain subject to the species via instinct, which limits the field of values and essences they can apperceive. And they are limited, on the other hand, by the physical Cosmos that bears them and that they rearrange. In brief, the creation of real beings is so successful that beings are at once free and yet made to work in a direction in which creation would encounter no obstacle. Provided they labor and exploit their faculties, they discover all that is indispensable to their existence: energy, material, fields of action of all kinds. To the point that they sometimes believe themselves to be true gods, children of chaos, the only conscious beings, the sole beings capable of judgment, choice, and projects. Creation is carried out so well that it remains invisible to the creatures. God guides beings without impelling them. And when beings, while benefiting from the resources of creation and using their language and brain to speak, declare that they have realized that God is only a myth, it is at that moment that God is satisfied and can proclaim his creation good.

This metaphysical fiction conserves a few of the solid conclusions of contemporary science. At least it has the advantage, like fast motion in cinema, of foregrounding the most fundamental lines and movements.

a. There is not only a universal reign of finality but a universal reign in a double sense. On one hand, all beings in the universe are centers of finalist action; they do not have a ready-made nature, but they make their nature according to an ideal that is itself modifiable. On the other hand, the general assemblage of the universe is arranged in such a way that the individual centers of finality act harmoniously without knowing or intending it. Put differently, active, dynamic, and individual finalities, capable of regulation, are based on a system that enables them (despite temporary conflicts) to converge and to adjust themselves. A "systematic" finality of the Kantian type is subjacent to regulative individual finalities. It need not be compatible en bloc with an equally en bloc determinism. It has to be compatible with the myriad of individual finalities. It is a fundamental Ruse of divine Reason, a ruse that has to be more profound than Leibniz's or Kant's finality (which is transcendentally compatible with determinism) or the Ruse of Hegelian reason, for it must envisage not only a one-piece determinism or a linear dialectic but also a history and a geography where accidents are real and yet where free beings can live and prosper. It is akin to the outline of a play written by a scenarist who has to leave a margin of freedom to his actors while maintaining the unity and beauty of his drama.

Despite the fundamental resemblance of all finalist activities, three great modes can be roughly distinguished: "mineral" activity, that of the individuals of physics and chemistry; organic activity with hereditary memory and instinct; and "conscious" activity with individual apperception of essences and values.[2] For each of these modes, the systematic plane that grounds them can be clearly discerned: chemical beings are situated in general tables with ready-made cells; the organisms proper, with much more subtlety, are equally distributed according to a general system of types; last, conscious activities also fulfill possibilities and obey *systems* of values. They are not absolutely free; they move within a structured "axiological space." Civilizations and cultures, as varied as they may be, are distributed according to their kinship in systematic tables. History does not cease to enlarge this table but also continues to fill it by conforming to cases that have already been made visible.

Yet in reality, these three modes cannot be isolated. Physical laws and moral laws appear absolutely opposed only in the deceptive perspective of a mechanistic and deterministic science that claims to absorb biology and no longer knows what to do with ideals and values. If continuity is reestablished in the universe of "organisms" in the broad sense, then the Kantian opposition between the obeyed law and the law that one represents to oneself will be reduced to a "modal" opposition; there are not *two* opposed types but *three* modes of laws, corresponding to the three modes of activity: laws of microscopic physics; laws of organisms subject to instinct; laws of consciousnesses that aim for values. And these three modes are not essentially different: it is always a matter of finalist obedience to a norm, not passive and deterministic obedience to a pure impetus. The norm is only more or less compulsory. An iron molecule has scarcely any choice. A bee that obeys instinct already has more. A human who apperceives an aesthetic or moral value has a great deal.

To these three modes of primary laws, all the secondary and statistical laws are opposed; the "secondary" laws also result in an order, but by pure equilibrium. It is these laws, in conjunction with the microphysical laws, that guarantee the order "of the starry sky over our heads." The laws of aggregates combine with both the biological or psychological laws and the primary physical laws to govern the general behavior of beings and restrict their freedom.

The three primary modes of finalist activity do not preclude unity, because these activities aim, if not for the same regions of the domain of essences and values, at least for regions that partially overlap, as if Logos itself lay behind the particular ideals. Instinct and intelligent consciousness often double each other. They can be vicarious or can complete each other: maternal love extends parental instinct; intelligence and instinct invent the same tools. Mixed or transitional modes are possible: crystallizable viruses act at once as molecules and as organisms; intelligent activities are often half-instinctive. The three modes envelop one another in the same being: humans simultaneously obey the laws of microscopic physics, statistical laws, organic laws, psychoorganic laws, and mental laws. Human development is at once organic, psychological, and spiritual. By the same token, it is quite pointless to follow in von Uexküll's footsteps and to distinguish for the organism a plane of formation (embryogenesis); a plane of functioning (physiology); and a plane of repair (regeneration).[3] Everything is imbricated: a plant continues to grow while functioning, and this is true of the majority of animals and humans.

b. We see that the emergence of novelty is produced everywhere in the universe, because each domain of activity forms itself according to its own ideal and because every association of domains plugs into new ideals. But there is no "emergence" in the special sense in which the theories of "emergent evolution" employ this term. The "new" is formed at each instant and everywhere, but there are no superposed layers in N. Hartman's sense, each introducing a characteristic "novum."

God, as the site of all ideals or as the universal ideal, does not cease creating through the medium of all beings. But there are no clearly distinct levels in the universe, each of which would be equivalent to a second, third, or fourth "creation" or "general emergence" superimposed on the previous one. We have already noted how much the conception of layers is contrary to the rigor of the discoveries of contemporary science, which revealed the "fibrous structure" of the universe, that is, the lines of individuality that cross time. But it is worthwhile to insist on it, because philosophy seems to have a great deal of trouble in disabusing itself of this notion of specific layers of reality, chronologically and logically superimposed.

The levels of emergence cited most often by authors are life, consciousness, and value. None can be maintained. The life and the primary consciousness, the life and the "autosubjectivity" of each organic form, are identical: they are inseparable from value or at least from the "normative ideal" in the broad sense. Undoubtedly a man's life, consciousness, and world of values are disproportionately more complex than an atom's, but metaphysically they are of the same kind. The true layers are constituted by the different regions of the transspatial. They have nothing to do with levels or layers that differ in their reality within our spatiotemporal universe. We have seen that the accumulation of elementary individualities into aggregates constitutes a Cosmos, which serves as a Terrain for more sophisticated beings; but this is merely a secondary phenomenon, ingeniously combined with the great, more fundamental fact of lines of individuality. Life does not rest on physical reality; consciousness does not rest on life, like oil on water. Rather, higher living and conscious beings "break through" the crowd of elementary individuals in the sense in which we say that a talented person "breaks though," "succeeds," and "makes it." On this point, Leibniz's metaphysical cosmology, which explains the emergence of levels through the hierarchy of monads that have reached a more or less elevated state, is in better agreement with contemporary science than the majority of more recent metaphysics.

To express this in the language of our metaphysical fiction, God prepares and calculates everything in advance; the possibility of higher beings already existed in the very nature of elementary beings. Without a doubt, human intelligence (or the values it "apperceives") is not virtually contained as a *predicatum* within the primitive atom, but it is possible because the primitive atom is already a domain of absolute survey. Microscopic individuals are not monads, substances that enclose in advance all their predicates, all that they will become (on this point, the conceptions of emergence are obviously more accurate than Leibniz's philosophy); but they are already centers of activity capable of becoming all that they desire if they have, as Butler says, "faith," and if they know how to set about it in the right way.

The theory of "layers" and of emergence distorts everything. We can discern the "emergentist" prejudice—combined with the influence of Hegel, for whom life emerges dialectically from matter and the spirit from life—even in philosophies that do not lay claim to it; and it explains some of their most questionable theses. The word emergence, *which, as a matter of fact, signifies nothing or signifies etymologically the opposite of what it is made to say, imparts (like the word* organism*) a good scientific conscience to those who glimpse the fact of cosmic finality and refuse to recognize it but do not seek to return to the old mechanistic materialism. The thesis of emergence is a sort of laicized, diluted, or inverted creationism. Alexander considers God as a final emergence, God or rather the quality of "deity." "That the universe is pregnant with such a quality we are speculatively assured,"[4] but deity is fated to remain ideal and can never become actual.[5] Space-time, not God, is the creator of the world; "in the strictest sense [God is] not a creator, but a creature."[6] Despite all of Alexander's metaphysical subtlety, we cannot help but find here a metaphysical transposition of the science of the nineteenth century and of the despotic reign of Spencerian evolutionism.*

c. We see thus that no logical incompatibility exists between human finality in the universe and the finalist assemblage of the universe as a whole.

N. Hartmann (who put forward a very popular thesis in philosophy) considered it necessary to choose: either to believe in human teleology or to believe in the teleology of nature. He drew his inspiration from Kant, who, according to him, conclusively refuted the thesis of an individual finalist activity *(Zwecktätigkeit),* leaving only the idea of an en bloc finality of the world, which can alone be reconciled with determinism *(Zweckmässigkeit).* The belief in a finality of nature results from an anthropomorphic illusion, "[which] subordinate[s] ontological to axiological points of view, . . . conceive[s] of the world-process at large as an actualization of what is valuable in itself."[7] For Hartmann, as for Alexander, the axiological categories are "emergent" relative to the ontological categories that presuppose them; they are "higher" but "weaker." The finality of nature is simply a myth: God, the conscious "Subject" of this finality of nature, is just a magnified man; one attributes to him "predestination" and "providence" on the model of finalist human activity. From this point of view, no essential difference exists between theism and pantheism: the teleology is the same, except that in the case of pantheism, the teleological system remains "up in the air." In a finalist nature (finalist "en bloc"), finite beings like humans are impotent; God or the "spirit of the world" achieves its goal over their heads. Between deism and morality—Hartmann reverses the Kantian thesis—there is no compatibility: "The metaphysical humanization of the Absolute is a moral annihilation of man."[8] Because human morality and finalist activity are facts of experience, it is the teleology of nature—a pure theory—that has to be abandoned. It is atheism and not deism that is the "postulate" of human morality and freedom.

We admitted in advance that some measure of truth remains in N. Hartmann's argumentation: every finalist conception of the world, every deism, presupposes anthropomorphism and mythology. But let us consider for a moment another thesis quite distinct from the first: the incompatibility of a finality of the world in its totality and the finalist activity of humans. It cannot be maintained. Hartman is no doubt envisioning systems like Leibniz's, in which creation according to the principle of the best leaves no meaningful room for human freedom, or like Spinozist pantheism, where humans are merely "modes" and are free only through a mystical identification with God. Perhaps he is envisioning Aristotle's immobile First Mover. But these systems do not genuinely attribute finalist activity to God. God does not in reality create, he lets his nature function: it is his necessity and not his finalist freedom that entails necessity for humans. How is a finalist "divine plan" incompatible with "missions" left up to the freedom of multiple agents? Pure spontaneity-freedom would be incompatible with the finality of Nature or God, but not work-freedom. Work aims for an ideal. The divine plan can thus constitute the system of Ideals. The "system" imposes limits on all freedoms, but first of all it constitutes them. The highest authority in a hierarchy issues orders and assignments, but it leaves a margin of freedom to its subordinates. The author of a scenario can simply transmit an outline to his interpreters; he does not necessarily transform them into marionettes. The fowler or the bird charmer counts on the instinct of birds, but he knows full well that this instinct does not function like a mechanism. His general plan succeeds by and large, but each bird acts with the freedom inherent to the thematism of instinct. That we can reduce a finalist action that is realizing an assigned mission to a disguised fatalism is just a dubious philosophical fantasy. An activity that aims for a goal is never absolutely mechanized. A goal is not a magnet. Only an infantile imagination can believe that the horse will indefinitely draw the cart if we hold in front of it, beyond its reach, a sack of oats: due to a lack of "confirmation," an internal inhibition would rapidly bring this contradictory "finality-based mechanism" to a halt. God relative to us is not like Grey Walter relative to Elsie and Elmer (artificial electronic turtles). These automata only exert a pseudo-finality based on mechanical self-regulation and feedback, which can be reduced to pure causality. Their apparent finality is borrowed entirely from their builder, who decides to direct them toward the light or toward warmth. By contrast, our ends are truly personal, although they are connected to universal finality through the play of thematic instincts. Within the limits of our instincts, we glimpse values, and it is the consciousness of values and senses—that is, consciousness tout court—that turns us into demigods.

d. Not only is there no incompatibility between conscious finality and the finalist assemblage of the universe but the one necessarily presupposes the other. The active, dynamic and "working" finality of living and conscious individuals presupposes a fundamental teleological order that renders this individual finality possible. Bosanquet and

L. J. Henderson in particular rightly insisted on the fitness of physical nature and of the primitive properties of a few fundamental bodies, which makes "stable, durable, and complex, both the living thing itself and the world around it."[9]

Carbonic acid and water have specific "ideal" properties (water especially, with its surface tension, its specific heat, its higher density compared to ice, etc.) that make the constancy of the environment and the mobilization of chemical elements possible. The three elements (hydrogen, oxygen, carbon) are perfectly suitable for the construction of complex molecular composites, stable yet capable of very energetic reactions and of equally complex and balanced phases or cycles. Their properties "make up a unique ensemble of properties each one of which is itself unique."[10]

Contemporary physics and chemistry can add much to Henderson's expositions. There is something dizzying about the complexity of a higher organism's conditions of physical existence. The anatomic complexity of the eye, which gave Darwin fever, is totally insignificant alongside the complexity of the whole of microphysical, physical, chemical, and physiological structures that enable the emission of light and vision in general. Bernal and Fowler's recent study on the structure of liquid water reveals the world of complications implied by the properties that make water an abnormal and exceptional liquid. Owing to the presence of two hydrogen nuclei, the water molecule presents positive and negative poles that are arranged in a calculable way by the wave mechanism and allow three different approaches to molecules. Water is a mixture of three types (type "tridymite," type "quartz," and type "cristobalite"). The structure of type 1, where the molecule is the most voluminous, resembles the structure of ice, as if it retained a "memory" of this structure, and is transformed into type 2 when the temperature rises.[11] To understand how an animal could drink when it feels thirst, we have to climb up to the fundamental nature of molecules, atoms, atomic components, space and time, quantum of action, coupling of electronic spins, and so forth. It is useless to say that organisms are adapted to the environment they discover (whatever it may be), because it is the *adaptability* and not the *adaptation* of higher organisms that has to be explained by the very nature of the environment, from which they are inseparable: "The inorganic, such as it is, imposes certain conditions upon the organic. Accordingly, we may say that the special characteristics of the inorganic are the fittest for those general characteristics of the organic which the general characteristics of the inorganic impose upon the organic."[12] On account of the entrenchment of the organic (of higher organisms) in the physical world, adaptation cannot have a unique direction; and there must be a reciprocal conformity. The more the advances of physics allow us to "track" the structural formation of the (exceptionally suitable) traits of the environment, the more it becomes clear that the very order of the physical system, which envelops the exceptional by relating it to its rule, is assembled in such a way as to make the life of higher organisms possible. "We are obliged to regard this collocation of properties as in some intelligible sense a preparation for the processes of planetary evolution."[13]

According to the opposite thesis of N. Hartmann and contemporary existentialists, a man who is thirsty and searches for water, who is hungry and searches for fruit to eat, performs a finalist action, but this human teleology is possible only in the absence of a teleology of nature. Humans can exert a free and finalist action only in a teleologically neutral or, in the technical sense of the term, "absurd" world. One should recognize that plausibility is not on the side of N. Hartmann's or the existentialists' thesis.

e. Of the three principal modes of finalist activity distinguished in the preceding, our metaphysical fiction amounts in sum to considering the third (conscious and rational activity) as more fundamental than the other two, because it is on the model of conscious human activity that we have, as Witnesses of God, seen the formation of the universe. It is well known that the greatest reproach constantly leveled at finalism is the objection of anthropomorphism. Without tackling the heart of the question, what is the actual value of this reproach?

Hume stated it compellingly in his *Dialogues Concerning Natural Religion.* Thought, says Philo, is after all only one of the powers or energies of nature whose effects are known but whose essence is incomprehensible. "In this little corner of the world alone, there are four principles: *reason, instinct, generation, vegetation.*"[14] The world resembles a living creature, an animal or a plant, perhaps more than it resembles a machine (Hume means here Paley's "machine presupposing the Almighty watchmaker"), "and when Cleanthes asks me what is the cause of my great vegetative or generative faculty, I am equally entitled to ask him the cause of his great reasoning principles. . . . But reason, in its internal fabric and structure, is really as little known to us as instinct or vegetation."[15]

We can easily transpose Hume's objection by applying it to our three modes of finality: why consider conscious, rational activity rather than organic activity or "mineral" activity as more fundamental? God is supposed to envelop at once the physical world, the organic world, and the rational world. Why conceive God or the Ground of everything as supreme Consciousness-Reason rather than as supreme Instinct or supreme Mineral? Because statistical determinism represents a kind of fourth mode, which no longer has any finalist aspect, why not even conceive God following ancient atomists, or materialists before quantum physics, as "supreme Aggregate"?

Hume's argument loses much of its force once it is transposed in this way. It is clearly impossible to admit, after the progress of contemporary biology and psychology, that consciousness is merely one "power" of nature *alongside* other powers, with a different and unknown essence: instinct, generation, or vegetation. These three "powers" are one and, furthermore, they cannot be separated from consciousness. As interested as psychologists are in finely distinguishing between the multiple stages of the birth of intelligence, or the passage from parental instinct to maternal love, or the various successive stages of awareness, they would be reluctant to speak of a distinct metaphysical principle for each of these stages (on the details of which they do not even agree). We have at any rate attempted to show that the general traits of unitary domains of survey are subjacent to the various modes of intellectual consciousness and instinct, and even microorganic or microphysical activity. It is thus no longer a matter of opposing supreme Mineral and supreme Reason: regardless of which mode is chosen as more fundamental than the others, God is always conceived on the model of a unitary-domain Agent.

But let us even accept Hume's argumentation as such and simply complete it with the additional hypothesis of a "Mineral God" and an "Aggregate of atoms" God. It is double edged. For if the Consciousness-God is not more justified or more explanatory than the Plant-God or the Mineral-God, the reverse is equally true, and "Vegetomorphism" or "Cristallomorphism"—if we can coin these expressions—is not more justified than Anthropomorphism or Logomorphism. In Hume's time, the belief in a Consciousness-God was very widespread. Among contemporary philosophers, and perhaps among contemporaries tout court, the majority who claim and believe they are atheists have a rather vague preference for an "Aggregate of atoms"-God or a blind Instinct-God. Hume's argument holds against them as much as it holds against the theists of his time. Why one *-morphism* rather than another?

If there is a logical indifference between these options, there will be no indifference once we examine each of the hypotheses more closely. The blind Instinct-God is highly questionable. It is today a legacy of romanticism, which, through Schelling and Schopenhauer, profoundly influenced the whole subsequent philosophical thought. Everything indicates that instinct as blind Nisus cannot be a primitive Fact. Instinct, like the drive, is a "means" of organic life. The instinct of reproduction, for example, is obviously relative to the total sense: "life of the species"; it is relative to a Logos that is segmented between several individual bearers; it is the "dynamic custodian" of the unity of the segmented cycle. The vegetative or generative instinct is blind insofar as it is dynamically subordinate to the goal to be achieved. Like every subordinate means, it sometimes functions on its own and mindlessly. But it is absurd to make this accidental mindlessness the very ground of reality. Venus or Shiva can be gods, but not God.

One has to be careful: the notion of a supreme Organism-God can be taken in two very dissimilar senses. If one considers the "organism" in its most general sense as a "unitary domain of activity and assemblage," Organism-God reverts then to Reason-God or Consciousness-God; and the expression can even be advantageous because it allows us to escape the lamentable confusion of the neural and cerebral auxiliary with primary consciousness, inherent to every "organism," and with the perceiving or fabricating consciousness in external circuit. A "Brain-God" is certainly not a more coherent concept than an "Organism-God." What has to be critiqued is thus the blind Nisus-God. The Organism-God in the other sense does not essentially differ from the Reason-God.

The Mineral-God has to be approached with caution as well. It can also be understood in two distinct senses. If one conceives God on the model of physical realities like elementary organisms or on the model of a general Plan(e) of the physical world, this conception will not at bottom differ from the conception of the Reason-God; or it will be easy to assimilate the two. But if one understands by the Mineral-God a kind of transposition of classical physics into the Absolute, one falls into a more palpable absurdity than that of the blind Instinct-God. Reality cannot be conceived on the model of aggregate phenomena, which are produced *within* reality. It is difficult to install in "God" in the ordinary sense of the term the equilibrium, the periodic oscillations around an equilibrium, the progress toward maximum entropy, the progress toward statistical order, or any phenomenon of macroscopic physics (condensation, rarefaction, sedimentation, fluctuation, etc.), and the doctrines that consider such phenomena as philosophically fundamental are generally classified as atheist. But the words have no bearing on the matter. To consider a statistical phenomenon, an aggregate phenomenon, as fundamental, as Absolute, is indeed to consider it as God. The true God of ancient atomists is the aggregate of atoms and their fortuitous combinations and not the subtle bodies that Epicureans bizarrely lodge in one canton of the universe. The concept of "God" is meaningful only as a propositional function; it is neither a proper name nor the equivalent—condensed into three letters—of a picturesque description. It can thus be said that this form of atheism, which posits the world of classical physics as Absolute, is merely a poor deism. It has an idea of God as naive and contradictory as the simplest anthropomorphism. It is moreover a disguised anthropomorphism, because it posits as the only conceivable possibility the state of the universe at the scale, if not of human beings, at least of an all-too-human physics.

*As Dilthey (*Weltanschauungslehre, *1911) and Leisegang (*Denkformen, *1951) demonstrated, this positivist and naturalist conception thinks it escapes the questions of origin by extending the line of ordinary physical processes to infinity. Characteristically, P. Laberenne*[16] *considers the thesis of a temporal origin of the universe, popularized by the discovery of an expanding universe, a great danger to the scientific conception of the world. In fact, this presumed origin—that is, the epoch in which the radius of the universe was at a minimum—would only date back some 10 billion years. Beyond this moment, one may then be tempted to believe either in a creation, like G. Lemaître, or in a state of the real that is completely different from the state of the universe of science. But* fortunately, *continues Laberenne, the physicist R. C. Tolman showed that because the stars and the galaxies are much older, we should instead admit a series of oscillations from the smallest radius to the largest. "The life of the universe would thus be composed of a succession of festoons, each of which* *would last for instance a hundred billion years; the average age of stars would extend over twenty festoons; the average age of galaxies over two thousand festoons; the present epoch (ten billion years) would correspond to the passing of some ten festoons." The origin of galaxies is thus so distant that it no longer seems to pose any metaphysical problem for Laberenne and Marcel Boll. More likely, "positivist naturalists" hope that astronomers and physicists will continue to imagine vaster cycles that indefinitely prolong the reign of the physical phenomena we know.*

It is nevertheless clear that aggregate phenomena, statistical determinism, and the oscillations and fluctuations cannot be fundamental phenomena. One might as well say that oceanography supplies the key to understanding the nature of a water molecule. Fortuitous or statistical fluctuations can produce nothing. To appear productive, they have to be "captured" by a consciousness lying in wait (Maxwell's demon or an inventor who is reaching for a solution) or, more generally, by an order of possibilities subjacent to the fluctuating phenomena.

Contrary to what is sometimes believed, the arithmetic of probabilities never bears on chance but on the structure subjacent to the fortuitous combinations. The chance of obtaining six in a die throw is "one-sixth" not because the "laws of chance" will it so but because geometrically the cube has six equal faces. As G. Matisse says, "the so-called laws of chance relate to something other than chance; they are statistical laws that can be applied to collective aggregates with a determined and known constitution."[17] Chance and statistics can only reveal a preestablished order; they cannot create it. As soon one presses it slightly, the conception of Chance-God or "aggregate phenomenon"-God leads to that of the Order-God, indiscernible from a Mathematician-God or a Reason-God. The hypothesis of natural selection, especially if it is considered in the abstract form of a pure mechanical sorting, in no way bypasses cosmic finalism. On the contrary, it amounts to shifting the whole weight of this finality to a kind of mathematical Order subjacent to the games of chance. Organisms are then produced in conformity with preestablished possibilities, which fix in advance, eternally, their conditions of existence according to the laws of a combinatory topology.

It is now time to tackle the heart of the problem and to account for the openly fictional character of our metaphysics. The internal assemblage of the universe is such that finalist activity reigns everywhere within it: all beings are domains of activity; all "agents" aim for an ideal or conform to it in one way or another. As such, it matters little whether we conceive God on the model of the human agent, the organic agent, or the mineral agent. For in any case we fall into a much more serious contradiction than any anthropomorphism. The contradiction is as follows. Suppose the universe is assembled by a transcendent God in such a way that finalist activity in its various modes can reign within it—is this assemblage a finalist activity? If we answer with no, this means that finality is not fundamental after all—that it is just a fact in the world and that there is no Logos or Sense of the world. If we answer with yes, we will be condemned to an irremediable infinite regress. God is to the world in its entirety what any agent in the world is to its unitary domain and to its ideal. But what is God's "ideal"? The fundamental unity of all the modes of finality makes the difficulty all the more palpable. If *every* finality presupposes agent, unitary domain of work, and ideal, does the finality *of* the world, that is, the fact that it is assembled in such a way as to render particular finalist activities possible, require in its turn agent, unitary domain, and ideal? God as Sense of senses or End of ends is thus no more intelligible than God as Cause of causes or Being of beings. In both cases, one is caught between an infinite regress and the negation of the concept one sought to raise to the second power—which seems to reduce the concept itself to a fantasy. Either the Sense of senses is senseless or we have to search for the sense of the sense of senses, and so forth. Although N. Hartman and existentialists are wrong to assert that human finality presupposes the nonfinality of nature, its teleological neutrality, it seems undeniable that the sum of finalities that constitute the totality of the world cannot be oriented toward an end—as though there were, after all, some measure of truth in the philosophy of the absurd.

The solution of Whitehead—with whom we have been so often in agreement—is unacceptable in this case. Whitehead splits God into an "Ultimate," which he dubs "Creativity," and a "God" who is "its primordial, non-temporal accident."[18] *Either we are dealing with an infinite regress triggered and poorly disguised under this non-Manichean dualism or, if Whitehead's nonultimate God is the equivalent of the "Ideal" of the world acting as a* lure, *we are faced with an incomplete solution.*

There is one and only one way of escaping the contradiction: to identify God not with a being, a sense, or an activity transcendent to the world but with the two poles of all finalist activities whose totality constitutes the world. God is thus supreme Agent as well as supreme Ideal; and "Creativity" cannot be distinct from a God who is at once and indissociably Agent and Ideal. Because the world is made up only of lines of activity, God is at once the world and yet distinct from the world, for the multiplicity of activities plays as we have seen the role of a kind of opposition, the role of a material, for each particular activity. It plays the role of a resistance to the effort of signifying information in the same way that an aggregate is composed only of individuals and yet is opposed to each.

Let us reconsider our table. It represents in principle the entire world, because it would suffice to enumerate all the cosmic activities one by one to obtain the totality of the real.

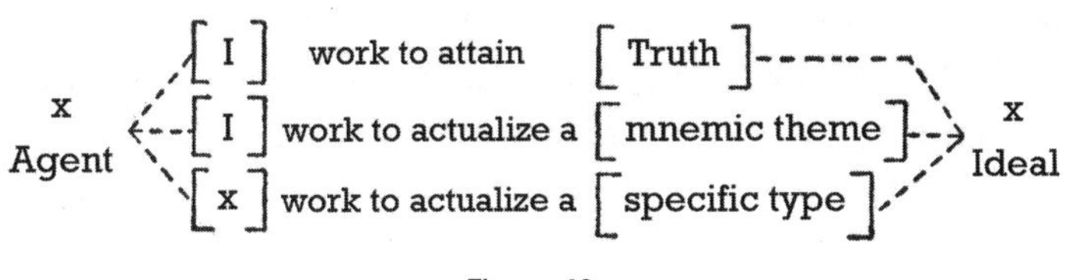

Figure 46.

The content of the left and right brackets designates something that is never completely finished or determined. There is something ungraspable about the Ideals; to grasp them is to automatically work according to them and therefore to incarnate them in one's particular line of existence and activity. The same holds for the agents. "I" only grasp myself in my act or, because this act enriches the "I," I only grasp myself as enriched "self," with habits and talents I speak of as of a foreign person. The "I" of the "I think" is ungraspable. As soon as the thinker speaks of it, he transforms himself into an object for a more distant "thinker," and so forth. But this is no less true of all activities and all works. Lequier's paradox, "make and, by making, make yourself," expresses this fact very accurately. To make oneself is to work before being. A being that is nothing more than an agent, that exists only insofar as it acts, cannot by definition ever grasp itself, because one can only grasp a being and not an activity, which is itself an activity of grasping. Our conscious "I" did not begin its existence by taking stock of itself; it acted by continuing an act, the formative embryonic act that itself continued a germinal act. There is an analogous law for the ideals and the values: a concrete goal is relative to an end, an end to an ideal, and a particular ideal to a universal ungraspable ideal. The labor that aims for the ideal leads to a work that, even and especially when it succeeds, interposes itself as a screen between the ideal and the agent.

Activity, work, which constitutes the whole "substance" of the world according to contemporary science, cannot be dissociated from its two poles, which are simultaneously intimate and transcendent to it. Yet can we speak of a single Agent pole and a single Ideal pole, despite the myriad centers of activity? Put differently, in all that is bracketed, is the Ungraspable homogenous, and can we speak of it as a single *x*? It seems so, because if we track any line of activity in the fibrous structure of the universe, we discover junctions with any other line. The ungraspable *x* that lies behind my "I," the Activity out of which the activity of my "I" has sprung, is also the ungraspable *x* of any other "I" or any agent living today. Reproduction by self-replication cannot be, as we have seen, a mechanical tracing; it presupposes an internal unity between two bifurcating lines at the moment they fork out, in the same way that the fusion of two lines during fertilization produces a single being who says "I" despite its two parents. G. Lemaître's hypothesis of the primordial atom may be true or false; but it proves at least that bifurcations in the life of the atom are not inconceivable. Even if the multiplication of microphysical individuals were impossible in space-time, this would not mean that an ungraspable unity of all individuals, of all "agents of matter" (according to the expression of H. Weyl and de Riezler), is impossible in the transspatial. Likewise and symmetrically, although the demonstration is more difficult and cannot rely on scientific probabilities, the singularization of the term "Ideal" seems justified by the fundamental unity of all Ideals. The mnemic themes are coordinated in vaster systems, insofar as memory and invention are indiscernible. Similar ideas are numerically the same ideas. Millions of individuals can have the same ideal. Despite the conflicts of values, the notion of a supreme Ideal is all the same less mythical than the image of a struggle between God and the Devil or between Ormuzd and Ahriman.

The cosmological continuity of existences, like the relative harmony of ideals, can thus pass for the expression of something more profound. If the x never begins to exist, and if it cannot grasp itself as an object, this is because it is God. It is God who exists in each of us as he subsists in each of the Ideals. There is, let us admit, no free being; there are only free activities. We must rectify this formula as follows: there is only one free *being,* God in us, and we exist solely by creating, that is, by working according to the order of the ideal, which is also God in the Ideals. Thus God does not in fact create free beings or free activities, which would detach themselves from him. This was the mythical element, the cause of antinomies, in our fiction as in every creationism. A free being cannot be created; it is "continued creation," that is, "continued God." The paradoxical existential dependence of the agent on its activity does not apply to God in us. Our soul forms itself by forming our body and the extensions of our body that our tools are. But the soul of our soul, to borrow the mystics' expression, never has to be made because it is eternal and makes time like everything else. As we survive the changes of the objects on which we work, as we can pass from one activity to another, though it is our activity that enables us to exist, so God survives the very changes of bodies and souls. Our soul perishes with our body, but the soul of our soul changes bodies and souls as we can change the object of our activity. Zeus's metamorphoses are the symbol of this truth: God takes hold of and abandons us as we can take hold of and abandon an ongoing work, though we cannot genuinely cease to act.

We can perfect our table and transform it into a diagram for symbolizing the bonds between God and the world. From left to right (as before), the polarity agent → ideal of activity. From top to bottom, its increasingly incarnated character (through the multiplicity of beings and their colonizations). The chainlets represent the lines of individualized activity. Above the horizontal line I, God as Agent and final Ideal from which all activities are suspended. Beneath this line, the world, comprising at once the already "natured" transspatial (between lines I and II) and spatiotemporal nature (beneath line II). Between the two horizontal lines, the "I" or the principles of individuality, on one hand, and the figured Ideals, on the other, are not spatiotemporal existents in the strict sense of the term. Our "I," for example, is eternal relative to a limited domain of time, and it enjoys a kind of ubiquity relative to a limited domain of space; "I" am God relative to my life, if I at least unify it through an ideal (Figure 47).

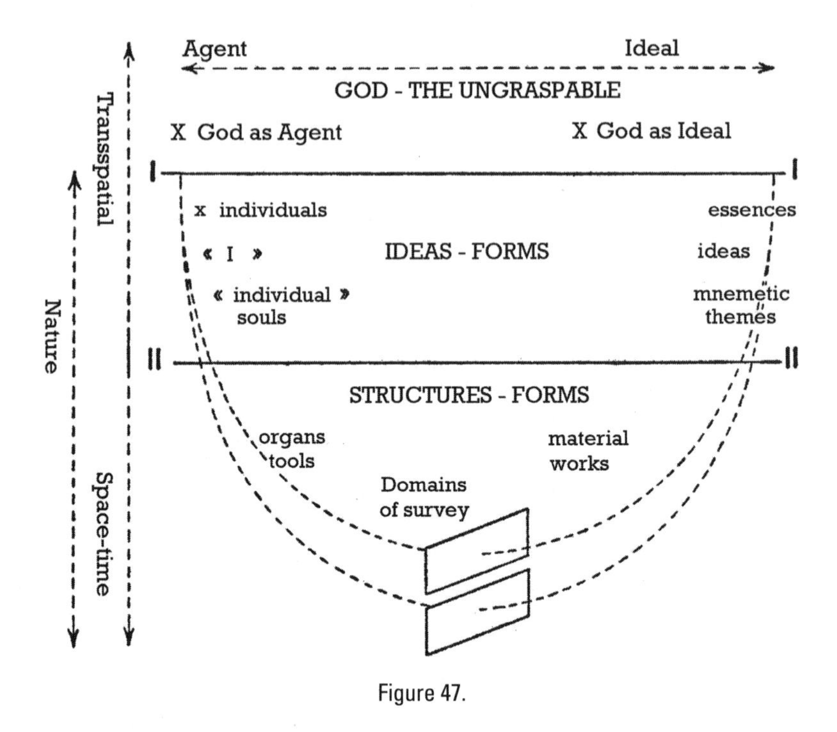

Figure 47.

Only beneath line II do the scenes of *observable* work begin, in which, on one hand, the agent appears as incarnated, as an assemblage of acting organs and tools, and in which, on the other, the work applies to concrete domains of survey through which the agent apperceives the Ideal that guides the transformation and improvements of the domain.

Lequier's paradox applies to what lies between the two horizontal lines: the "I" makes itself by making. Here author and work are inseparable. It is true as much for the embryo that forms and individualizes itself according to the mnemic potential of its species as for the artist who modifies himself and enriches his soul through his own works. That the blacksmith can take hold of or abandon his material work is merely a secondary phenomenon, possible only beneath line II. This phenomenon must not conceal the fundamental inseparability of the actor and the work, and it must be conflated even less—according to the crude confusion of vulgar creationism—with the transcendence of God relative to all the individuals whose being and most intimate activity he is.

No opposition figured on the diagram is absolute. The divine unity leaves no border impermeable. The region of ideas-forms is Nature relative to God, but it is God relative to space-time. Space-time itself is only constituted with actions, and in this sense, it is not opposed to the transspatial as a sort of opaque and rebellious matter. It is only opposed to the transspatial as a purely statistical average of the myriad activities that constitute it.

At the "level" of God himself, not only Agent and Ideal are inseparable, but the duality of the two poles is probably not real. God as Agent does not differ from God as Ideal. God has no faculties, no attributes, no distance from himself, no nature, because he is all that constitutes nature. Even for creatures, there never exists any abrupt separation between the pole Agent and the pole Ideal. There is never any pure agent or any pure target-object. Every being is simultaneously creator and creature. Reciprocally, a target-idea is always individualized: it is identified with the subject that constructs it, and it becomes active for this subject. The first incarnations of the active *x* of the organism are at once its works and its substitutes; they are themselves active. A theme of invention is invented, but it invents in its turn. The soul is formed but is also forming. The body is at once a work of art and a living tool, capable of forming pure bodies, that is, machines. A recollection is simultaneously an ideal norm that directs our groping toward it and an intimate auxiliary of the "I," an active habit and an "other I."

Subjacent divine plan(e) and finalist activities are not dissociated. God, mythically isolated, evokes the image of an artisan, like the ones that can be observed in nature, and whose finalist activity is superadded to all the others without explaining anything. But God cannot be isolated from the World. His finality is not superadded to finalities; it is their total Sense.

The tendencies of contemporary philosophy can be summed up with two statements: (1) "the world is absurd" and (2) "God is dead; but God has just been born, *it is I*." Not everything is false in these two statements, but the minor element of truth that they contain is totally perverted. The world is not absurd: it is made up uniquely of finalist and senseful activities. I am not God, not because God is "other than I," but because he is all of the other "Is" and their general sense. I stand against the background of the whole organizing nature that bears me. I concede the reference to Logos and Sense by speaking of absurdity. But it is equally true that, as divine and free Agent, I am *Natura naturans*, and I have the right to consider as neutral the whole *Natura naturata* up to the emergence of my freedom. I am even capable of inventing new values that I discover in God but that are equally in myself, to the extent that God and "I" are one. By virture of the purely statistical laws, nature as the reign of the multiple is what must be subjected to the norms of various values, what gives work to my finalist effort toward Sense. "Every logic presupposes error and every morality presupposes immorality."[19]

Nevertheless, this does not imply that error, immorality, and absurdity represent a kind of primordial ground, more primordial than God himself. The statement "God is absurd" is meaningless. God does not rest on an "other" that would be the absurd or non-sense. And he is not, like each of us considered separately, in combat with the reign of the multiple, for he is everything. We can turn the reproach of mythology against those who believe in a *Grund*. This *Grund* can only be a vain human fantasy, a residue of the mythological vision of Chaos and the Abyss, or a vague social memory of marshy lands before culture.

God is not the Agent of agents, their fabricator; he is the Agent who is in all agents. His freedom and his science do not contradict my own, for he is my freedom and my science. His eternity does not contradict my time, for time is not time—that is, something more than a pure multiplicity of instants that do not know one another—save through the eternity that animates it.

Last, the idea of God as Ideal and as Agent is not in contradiction with our mediocrities, our faults, our miseries, and our sufferings, which are also his own. The existence of negative values (ugliness, falsity, injustice, weakness, hatred, evil) has always been invoked as an objection against both pantheism or positive mysticism and finalism. But just as it is necessary not to confuse dark vision and null vision, so it is necessary not to confuse negative value and absence of every axiology. The philosophy that establishes the reality of finalism has no claim to being a theodicy.

Humans are too quick to harp on negative values and to claim, like one contemporary writer, that "India stinks of the devil as a urinal stinks of urine"—India, or the world of insects, or the world of reptiles, or the equatorial forest, or Behemoth and Leviathan. But of course Europe too stinks of the devil for Hindus with a delicate sense of smell, and Behemoth has to find the monstrous man. The gods of others easily become devils. Let us not slip into metaphysical and religious provincialism. We are too easily akin to those devout believers who imagine that God inhabits their temple or their little pious friary while the "World" is the kingdom of Satan. "God" is not synonymous with perfection; or at least, his perfection resides as much in variegation and opulence as in purity and harmony. It lies in the variety of dissonant accords as much as in the perfect accord. Improvidence, accident, chance, and misfortune can form part of the providential essence of a world in which divine freedom chooses to proliferate into myriad freedoms and finalities.

Summary

Our method has consisted in seeking isomorphisms between facts without fretting about traditional classifications.

After stressing the contradictory character of the negation of *every* finality and the impossibility of conceiving the finalist activity of conscious humans without linking it to an organic finalist world, we attempted to show that to describe the facts, we have to introduce a fundamental distinction between unitary domains of action and systems that are only tied by step-by-step actions. Causality without finality reigns exclusively in these systems. Unitary activity, that is, authentic activity, is finalist. And because the universe is nothing more than the ensemble of such activities, finality is universal. Causality proper is merely a derived mode, derived from the multiplicity of "actings."

Embryos and brains are typical examples of unitary domains. It is impossible to understand their mode of activity unless we ascribe to them an "absolute survey" that implicates a metaphysical "dimension" altogether different from the geometric dimensions of space-time. For objective observation, this absolute survey is translated by equipotentiality.

Yet all unitary domains are of the same general type; they are at once spatiotemporal and transspatiotemporal. They are "true forms" through the informing, dynamically effective activity of an Agent that aims for an Ideal.

The antifinalist prejudice that predominates in the mind of contemporary scientists is merely a remnant of the extended reign of macroscopic physics. It results from the transposition into metaphysical dogma of what is only true of the secondary order of the laws of interaction in a multiplicity of true individuals. It recalls the error that physicists once committed when they imagined the atom on the model of a planetary system with trajectories regulated by a play of incrementally established equilibria.

In all likelihood, and despite the crucial role that secondary laws play, the universe as a whole is essentially of the same type as a unitary domain. It is in any case inadmissible that it be a pure multiplicity, a kind of absolute Aggregate. Numerous clues suggest that individual finalities are subordinated to a total Finality or Sense. All finalist individual activities are isomorphic. All of them comprise "Agent → Work → Ideal." Strictly speaking, total finality is not isomorphic to individual finalities, not because it differs from them, but rather because it constitutes them.

Translator's Afterword

The Idea of the End

ALYOSHA EDLEBI

AN ANCIENT IMPASSE

For a long time, philosophy has lived off a cry that still shadows its planetary movement and imbues its concepts with a wild power: *Being thinks*. Parmenides bequeathed this formula to the Western tradition, along with the effort to instate a rigorous discipline that can draw its consequences and denounce its simulacra. We must not forget that, as masked or aphoristic as it might seem today, the cry has articulated philosophy's major imperative since antiquity: to think Being outside-of-thought, to conceive and situate the unthought of Being.[1]

Now, after two millennia of philosophical clinging to this cry, it would be hard to believe that Parmenides misunderstood thought's primordial datum. The given does not, in reality, stand for a specific idea or truth; it is nothing other than the existence of the unthought: that to which, in its upsurge, thought indeterminately hearkens and which it cannot seal in a concept or hand down in a doctrine.

Thought knows itself through its own production. So much so that it is always forced to abandon an unthought in its objects. *It is this forsaken dimension that conditions its advent into Being*. We might then rightly wonder: what does thought conceal in sheltering immemorial remnants of this type? What is the inconceivable that is left in every idea, the imperceptible in every perception, the unsayable in all of language?

One of philosophy's perennial tasks is to dispel the well-known theological image: the ineffable and the imperceptible make up the limits of thought. By doing so, philosophy shows that the unthought has no positive content and demystifies itself at the very foundation of perception and language. *Anoia* is less an impervious obstacle to thought than its *archē*-ground, the *anhypothesis* of every *noēsis*.

In the Greco-Roman world, to which we owe the motif of an essential proximity between grounding and negation, philosophy envisioned this nameless zone as the Absolute. Our time has hardly strayed beyond the classical intuition. While *absolutus* habitually denotes the Outside-of-thought, its other sense does not lag far behind: thoughtlessness, the absent ground *of* thinking.

Because thought can lay open its own inexistence, because it can render its own absence an *existence* and dwell enduringly there, it is at odds with every mechanism of foundation, with every will to ground.

In Hegel's wake, the aporia that flashes up before us here concerns the modalities of the negative: How can *in*existence exist as such? How can thought think its own absence in the opening of Being, its own interruption, its vacant place? To unravel this impasse, I would like to concentrate on a category that has exercised a strangely powerful influence over metaphysics and the paradigms of history—namely, *telos,* the End.

PARADOXES: FRACTURE AND WORLD

From the outset, Being is plagued by a fracture it has no power to elude. Pure, nonrelational Being constitutes the disparity between determinate Being and non-Being, just as a thing consists of the singular noncoincidence of something and *its* nothing, *to pragma* and its subtraction. This minimal fracture subtends the Idea of genesis.

Let's probe one of its repercussions: the manifestation of worlds. How is a world created? If we restore to the fracture its efficacy, we see that a thing cannot become itself purely and simply, it cannot nullify its intrinsic absence, its nothing. By dint of this impossibility, beings—from the tick to the brain—burst forth and are snatched out of their isolation: they are worldly. At the limit, we rediscover a venerable paradox that lurks in the archives of metaphysics: *a world does not arise from the order of the possible, from a transcendent determinability; on the contrary, it is born of a regime of the impossible immanent to each being.*

Across this regime, absence undergoes an individuation that bounds it up with actual entities. It is dislocated from total non-Being, in the same way that shadows are told apart from darkness: *beings cast ontological shadows of their own.* So every essence is hurled through its absence into a tremendous realm of *inessences* that gnaw away at it. "To have a world" means to maintain an alliance with the inessential, the negligible, the contingent. No doubt it is not by chance that those animals and humans who are cloistered altogether in their essence appear to us worldless, almost beyond redemption. Herein lies the secret ordeal of philosophy: to hold fast to the inessential.

In keeping with this rehabilitation of the negative, a whole tradition has brought attention to a second paradox; it bears on the lineaments of the End implicit in Aristotle's syntagma *to hou heneka*. "Having an end" should not be deemed a deficiency but a prodigious, even incomparable power: *finite beings wield the one faculty that the infinite lacks—the power to grasp their own non-being.*

The vitality of these paradoxes is so little in doubt that even their far-off echoes have not been entirely muffled. Let's consider a concept that has undergone such fervent renewals in modern philosophy that its import and its density are now obscure: I have in mind *conatus*, striving, tendency.

If the deconnection of *to on* and *to ouk on* is incomplete, if there exists—despite the Greeks' lively objections—an aporetic Being of non-Being, then desire cannot be assimilated to a simple *conatus*: a desire to persevere in being and to desire this perseverance. Because the "fractured" thing moves in the empty difference of self from self, its becoming will be identical to the being of its fracture. In this respect, to persevere in being means to strive to grasp an absence that *is*, a discontinuity which varies ceaselessly. Everywhere, *conatus* qualifies one force above all: *the ever-renewed desire to embody, to localize, inexistence.*

FIGURA, SIGNUM, SPECTACULUM

At its roots, Being is thus a noncoincidence that orients the polarity of objects and worlds. As to thinking, it seeks the disarticulations of the real, the discontinuities pulsing within the empirical, preserving them in the realm of the Absolute.

What is at issue in this process is a transfiguration of the fundamental character of time and its experience. The time of manifestation is the time a being demands to expose itself to itself, to give itself its own non-being, to witness its own ungroundedness. Time, it might be said, is a limitless generation, not of bodies and ideas, but of the modes of intelligibility of inexistence. *Through their becoming, beings seize the being of their absence.*

At this point, care must be taken to avoid confusion between two key notions: *figure,* which pertains to the ontical plane, and *sign* whose sphere is the ontological. A figure is the becoming-present of a thing to its world, as much as the presentation of inessences, of their bodies and changes, to the essence they so wholly englobe. Yet in each figure inheres a sign that calls up the fracture at the source of figuration; it dissolves in this way the reciprocity of world and thing. The sign is precisely the intuition of inexistence: it neither presents the thing itself in its original status nor decomposes and withdraws it, but rather presents its nonpresentation, the suspension of its visibility.

It is fitting that the figures which have deserted or simply lost their proper signs should be delivered over to the spectacle, as we too often see today. In this context, *the spectacle is the integral visibility of the now that strips the figure of its blank face, detaching it from its non-being.*

(One of the most insidious strategies of control in modern societies consists in depriving human beings, not of their presence, but of their absence, not of their visibility, but of their concealment, in other words: of the slightest signs that betray their estrangement from the "world of the now.")

IMMANENCE AT THE END-OF-BEING

Being has a peculiar end, but this end is not to be mistaken for simple non-Being, a founding and inert absence. In an initial approach, the end-of-Being throws a singular light onto the amphibolical situation of Being *within* thought: on one hand, Being strains towards its end when its movement reveals to it its partition from non-Being; yet, on the other, the end-of-Being traces a zone where the *a priori* disjunction between the one and the multiple, identity and difference, necessity and contingency, is weakened or vanishes.

Beyond the event of existence and its opening to sense, thought gives shape to an enigmatic dimension lodged at the end-of-Being: *the im-possible.*

This is why paradox operates embryonically within philosophy, whose dialectical vocation it announces. *That it is possible to think the impossible, to conceive of im-possibility in itself, that human beings can gain access to the outside-of-Being without their thought falling into nothingness, implicates a radical rupture with the intuition at the heart of Western metaphysics and praxis: it is not Being which thinks itself in concepts and propositions but thought, in its very detachment from Being.*[2]

On that account, neither nature nor history but only thought can be answerable—at least at first blush—to a *telos* turned on its head. This End, whose archetype crops up in filigree throughout Greek speculations on the One, is absolute and inassignable immanence: the immanence of thought to the outside-of-Being.

(In light of Nietzsche's critical diagnosis and our preceding remarks, nihilism may now be defined as the conflation of the outside-of-Being with the Nothing, the abolition of *anoia,* of the unthought.)

THE TORSIONS OF THOUGHT

If thought is in solidarity neither with Being nor with its negation and if this twofold estrangement articulates the foundation and abyss of its power, then its genealogy has to be intimately reconstituted against the backdrop of decisive aporias in the ontological tradition, most notably those of the im-possible and the in-existent.

One of the vertices of this genealogy is the locus of thought in language, the ending of the concept in the sign, of *noēsis* in *logos*. Here language is displaced, ontologically and ethically, in its function and in the problematics it prescribes for philosophy: instead of communicating the differentiated content of the Idea or constituting the pure medium of its taking place, language transmits and simultaneously bears witness to that which remains once thought has addressed itself to itself. Language is the *final* seal that heralds the human fulfillment of thought, attesting that there is, as such, *nothing* left to communicate.

With this in mind, we can advance a provisional hypothesis: the empirical place of thought is the inadequation between gesture and language. Impressions activate the movement of ideas, which signs embody and express. Yet the movement is possible only if the sensible determination (gesture) can emit an indeterminacy (thinking) that is not immediately captured by a language. In short, human beings are those living beings who have absolved themselves of their gestures so as to carry out the ceremonial of language, the accordance of sign and sense, from which their world issues. By this token language voids and renews gesture insofar as it signals—indeed, *is*—the mute ending of thought, the cipher of its having-been.

So, ontologically, humanity's alienation from language and history amounts less to an experience of pure discord than to an "untimely" testament to its status: the speaking animal, canceling its own power of re-presentation, drifts outside-of-Being, in the lawlessness of thought.[3] Here at last, the ancient name-of-man (*zōon logon echōn*) no longer alludes to the living animal endowed with *logos* but to one who uniquely *possesses* the absence of sense.

In alienation, history ceases to signify the unconditional presupposition of humanity, as in aphonia, language no longer anticipates thought, annulling its contingency. Be that as it may, it is a poor deduction to conclude that the uprooting of thought from the historical sphere consigns it to some immaterial time, through which the matter of History acquires a consistent and legible figure. This segmentation is illusory: thinking recapitulates itself *qua* historico-linguistic; it leaps abruptly from the stratified folds of history to its fringes, just as it is interwoven with *logos* and yet indelibly aphonic.

EXHAUSTION IS A THEORY OF PASSIONS

What befalls Being at its end, humans when they take leave of history and language? Solely in these figures of departure does it become possible to touch on the passions.

At stake here is a doxa that haunts philosophy from its inception: every *pathos* falls back on a *logos,* its latent presupposition. If our delineation of the ontology of the End has hit the mark, then the correspondence between language and passion enjoins us to resituate them in a more original sphere: *pathos is the exhaustion of logos.*

What is the crux of this proposition? Let's recall that exhaustion concerns the divorce of being and becoming, the tension between existence and power; it sets into play the depletion of the possible, grasping power at the instant of its ultimate liquidation. Exhaustion is, in a certain sense, the vital time of thought, in which the being of the impossible, the being that can no longer become, springs to the fore.

On the plane of passions, to exhaust language does not in the least mean to abolish it but to sunder it from any utterance, so that the fracture between expression and the inexpressible may be mended. How else are we to account for that affection whose curious force has never ceased to animate the ethics and hallucinatory drama of humanity: joy?

Joy is the state in which we wholly possess language but, having exhausted its tonalities, no longer have a use for it; this is why rejoicing is wordless, why we have no vernacular for beatitude. (Hence Augustine's sharp distinction between *fruitio* and *usus,* which finds its unexpected rationale amid the passions of the exhausted.)

CODA: ETHOS

Have we truly grasped and taken the measure of what shall forever escape us? Have we registered the fading plea for an ethics that would be neither a piety nor a solitude? Only if philosophy is capable—and it is by nature—of inaugurating an *anoialogy,* of folding itself into a science of the Unthought, will it absolve itself of its negative ground and embrace its unease before the Absolute; and humans, ravaged by their passions, will exit the economy of Being that has entranced them with its chaos to be absorbed into the aurora of thought.

But can thought entrust itself to itself, to its anchoring outside of Being? Can *noēsis* found, without any other authority than the peril that such an obsolete gesture entails, an *ēthos*? And might this wager release the language attuned to the most demanding and human of vocations: *the thought-of-thought?*

Notes

INTRODUCTION

"La philosophie a le devoir d'éviter le snobisme, mais aussi de ne pas exagérer la timidité."

1 Raymond Ruyer, "L'esprit philosophique," *Revue philosophique* 1 (2013): 9; originally published in *Orientation. Recueil de conférences au Centre universitaire de l'Oflag XVII A* (Paris: Éditions de Champagne, 1946).

2 Raymond Ruyer, "The Vital Domain of Animals and the Religious World of Man," *Diogenes* 5, no. 18 (1957): 35–46; Ruyer, "The Status of the Future and the Invisible World," trans. S. Walker, *Diogenes* 28, no. 109 (1980): 37–53; Ruyer, "Dialectic Aspects of Belief," trans. S. J. Greenleaves, *Diogenes* 15, no. 60 (1967): 64–79; Ruyer, "The Mystery of Reproduction and the Limits of Automatism," *Diogenes* 12, no. 48 (1964): 53–69; Ruyer, "There Is No Subconscious: Embryogenesis and Memory," trans. S. Walker, *Diogenes* 36, no. 142 (1988): 24–46.

3 Fabrice Colonna, "Présentation," *Les Études philosophiques* 1 (2007): 1. This special issue on Ruyer includes essays by Ruyer, Colonna, André Conrad, Renaud Barbaras, and François Brémondy. A second special issue of the same journal focusing on Ruyer and science has recently appeared (2013) and includes essays by Colonna, Conrad, Georges Chapouthier, and Denis Forest.

4 Fabrice Colonna, *Ruyer* (Paris: Les Belles Lettres, 2007), 53.

5 Ibid.

6 Ibid.

7 Raymond Ruyer, *Éléments de psycho-biologie* (Paris: Presses Universitaires de France, 1946), 6.

8 Ibid., 8.

9 Ibid., 13.

10 Ibid., 9.

11 Ibid., 8.

12 Ibid., 10.

13 Ibid., 14.

14 Ibid., 14–15.

15 Colonna, *Ruyer,* 61.

16 Ibid., 85.
17 Raymond Ruyer, "Raymond Ruyer par lui-même," *Les Études philosophiques* 1 (2007): 6.
18 Raymond Ruyer, *Néofinalisme* (Paris: Presses Universitaires de France, 1952), 64–65.
19 Colonna, *Ruyer,* 85.
20 See Humberto Maturana and Francisco Varela, *Autopoiesis and Cognition: The Realization of the Living,* ed. R. Cohen and M. Wartofsky (Dordecht: D. Riedel, 1973).
21 See Ruyer, *La Cybernétique et l'origine de l'information* (Paris: Flammarion, 1954), esp. chapters 2 and 5. Ruyer's account of how consciousness frames information forms the basis for my account of the "framing function" in new media art in *New Philosophy for New Media* (Cambridge, Mass.: MIT Press, 2004), chapter 3.
22 Ruyer, "Raymond Ruyer par lui-même," 9.
23 Ibid., 9–10.
24 See the conclusion of Gilles Deleuze and Félix Guattari, *What Is Philosophy?,* trans. Hugh Tomlinson and Graham Burchell (New York: Columbia University Press, 1991).
25 Maurice Merleau-Ponty, *Nature: Course Notes from the Collège de France,* trans. R. Vallier (Evanston, Ill.: Northwestern University Press, 2003).
26 Renaud Barbaras, *Introduction à une phénoménologie de la vie* (Paris: Vrin, 2008), 157–81; Roger Chambon, *Le Monde Comme Perception et Réalité* (Paris: Vrin, 1974), chapters 6–8.
27 Paul Bains, "Subjectless Subjectivities," *Canadian Review of Comparative Literature* 24, no. 3 (1997): 511–26; Brian Massumi, *Parables for the Virtual* (Durham, N.C.: Duke University Press, 2002); Elizabeth Grosz, "Deleuze, Ruyer, and Becoming-Brain: The Music of Life's Temporality," *Parrhesia* 15 (2012): 1–13.
28 Mark B. N. Hansen, *Feed-Forward: On the Future of 21st Century Media* (Chicago: University of Chicago Press, 2014).

1. THE AXIOLOGICAL COGITO

1 [Throughout, translator's notes are in square brackets.]
2 Samuel Butler, *The Notebooks of Samuel Butler* (New York: E. P. Dutton, 1917), 337.
3 Wolfgang Köhler, *Gestalt Psychology: An Introduction to New Concepts in Modern Psychology* (New York: Liveright, 1992), chapters 1 and 8.

4 Ibid., 4.

5 Jules Lequier, *Recherche d'une première vérité* (Paris: Armand Colin, 1924), 138, 139, 141.

6 Charles Renouvier, *Dilemmes de la métaphysique pure* (Paris: F. Alcan, 1927), 172. Cf. also Renouvier's note to part 4 of *Recherche d'une première vérité,* 134. Cf. J. Wahl, *Jules Lequier* (Genève: Traits, 1948).

7 We can add W. Stern's fundamental argument from *Wertphilosophie,* vol. 3 (Leipzig, Germany: Barth, 1924), chapter 1, which he dubs the "axiological a priori."

2. DESCRIPTION OF FINALIST ACTIVITY

1 Arthur Eddington, *New Pathways in Science* (Cambridge: Cambridge University Press, 2012), 90. Cf. also Eddington, *The Nature of the Physical World* (Cambridge: Cambridge University Press, 2012), 341.

2 We showed this elsewhere: Raymond Ruyer, *Le monde des valeurs: études systématiques* (Paris: Aubier, 1948), chapter 9, 149–52, and Ruyer, "Métaphysique de travail," *Revue de métaphysique* 1 (1948): 26–54, 2 (1948): 190–215.

3 Cf. Charles Kay Ogben and Ivory Armstrong Richards, *The Meaning of Meaning: A Study of the Influence of Language upon Thought and of the Science of Symbolism* (London: Routledge, 1994), in particular, chapters 8 and 9, where Ogben and Richards cite and discuss twenty-three different definitions of meaning.

3. FINALIST ACTIVITY AND ORGANIC LIFE

1 Otto Friedrich Bollnow, *Teildruck aus: Systematische Philosophie, Vol. 2, Auflage,* ed. Nicolai Hartmann (Stuttgart, Germany: Kohlhammer, 1942).

2 The unity of organogenesis and instinct has become a commonplace; it has been underscored by a number of authors: Bergson, P. Vignon, MacDougall, Pierre Jean, Bleuler, Buytendjik, etc.

3 André Leroi-Gourhan, *Milieu et techniques* (Paris: Albin Michel, 1945), 409.

4 Ibid., 472.

5 Joseph Henry Woodger, *Biological Principles: A Critical Study* (London: Routledge and Kegan Paul, 1967), 436.

6 Jacques Bénigne Bossuet, *Connaissance de Dieu et de soi-meme* (Paris: Hachette, 1879), 4:2.

7 Cf. Henri Busson, *La religion des classiques: 1660–1685* (Paris: Presses Universitaires de Paris, 1948), 140.
8 Nicolas Malebranche, *The Search for Truth* (Cambridge: Cambridge University Press, 1997), 98.

4. THE CONTRADICTIONS OF BIOLOGICAL ANTIFINALISM

1 [In English in the original.]
2 Cyril Dean Darlington, *The Evolution of Genetic Systems* (Cambridge: Cambridge University Press, 1939).
3 [In English in the original.]
4 Kurt Koffka, *Principles of Gestalt Psychology* (New York: Harcourt, Brace, 1935), 10ff.
5 Wolfgang Köhler, "On the Problem of Regulation," in *Selected Papers of Wolfgang Köhler*, trans. Mary Henle and Erich Goldmeier, 305–26 (New York: Liveright, 1971).
6 Cf. Raymond Ruyer, *Éléments de psycho-biologie* (Paris: Presses Universitaires de France, 1946), 87ff.
7 Hugh Bamford Cott, *Adaptive Colorations in Animals* (London: Methusen, 1940). Figures 7–15 are from Cott.
8 Alfred Tylor, *Colorations in Animals and Plants* (London: Alabaster Passmore, 1886).
9 D'Arcy Wentworth Thompson, *On Growth and Form* (Cambridge: Cambridge University Press, 1940).
10 Cited by Julian Huxley, *Evolution* (New York: Harper, 1943), 414.
11 The camouflage of the eye, which lies at the point of convergence of the simulated leaf's pseudo-veins, is particularly worth noting.
12 Ellen Mary Stephenson and Charles Samuel Stewart-Evison, *Animal Camouflage* (Edinburgh: A&C Black, 1955), 71–72.
13 Robert Hardouin, *Le mimétisme animal* (Paris: Presses Universitaries de France, 1946).

5. FINALIST ACTIVITY AND THE NERVOUS SYSTEM

1 Ruyer, *Élément de psycho-biologie.*
2 They betray themselves with the use of military metaphors. A French biologist recently said that "finalists now have only some islands of resistance. . . . They are no longer dangerous, but they should nevertheless be driven out." He did not specify whether he intended to use grenades or tear gas.

3 Andrée Tétry, *Les outils chez les êtres vivants* (Paris: Gallimard, 1948), 312. Cf. Lucien Cuénot, *Invention et finalité en biologie* (Paris: Flammarion, 1941).
4 Tétry, *Les outils chez les êtres vivants,* 319.
5 Cf. Stephenson and Stewart-Evison, *Animal Camouflage,* chapter 10.
6 [In English in the original.]
7 Cf. Jennings, Métalnikov, Mast and Pusch, Piéron.
8 Von Monakow and Mourgue, *Introduction biologique à l'étude de la neurologie et de la psychopathologie: intégration et désintégration de la fonction* (Paris: F. Alcan, 1928).
9 J. R. Kantor, for example, pursues the path of "abto-cerebralism" so far that it is no longer clear what purpose the brain serves in the organism. Kantor, *Problems of Physiological Psychology* (Bloomington, Ind.: Principia Press, 1947). L. P. Jacks speaks, for his part, of the "brain-myth." Jacks, "The Brain-Myth," *The Hibbert Journal* 41 (1943).
10 [English in the original.]
11 Fire detectors function in case of a *sudden* rise in temperature.
12 [In English in the original.]
13 Cf. Harold Locke Hazen, O. R. Schurig, and Murray F. Gardner, *The MIT Network Analyser: Design and Application to Power System Problems* (Cambridge, Mass.: MIT Department of Electrical Engineering, 1931).
14 Cf. N. Wiener, *Cybernetics: Or, Control and Communication in the Animal and the Machine* (Boston: MIT Press, 1948), and esp. E. C. Berkeley, *Giant Brains or Machines That Think* (New York: Wiles, 1949); cf. also volumes 1–3 of *Bulletin of Mathematical Biophysics,* in particular the articles of N. Roshevsky. E. C. Berkeley, *Giant Brains,* provides an assembly model for building an elementary electronic calculator.
15 Ruyer, *Élément de psycho-biologie,* chapter 8.
16 Berkeley, *Giant Brains,* 12.
17 René Descartes, *Discourse on Method* (Indianapolis, Ind.: Hackett, 2001), 46, translation modified.

6. THE BRAIN AND THE EMBRYO

1 Karl Spencer Lashley, *Brain Mechanisms and Intelligence* (New York: Hafner, 1964).
2 [In English in the original.]
3 [In English in the original.]

4 See Carlyle Ferdinand Jacobsen, *Studies of Cerebral Function in Primates* (Baltimore: Johns Hopkins University Press, 1936).
5 Cf. Clifford Thomas Morgan, *Physiological Psychology* (New York: McGraw-Hill, 1950).
6 Cf. Köhler, Buytendijk, Guillaume, Bierens de Haan, etc.
7 L. Verlaine, "Psychologie animale et psychologie humaine," *Recherches philosophiques* 2 (1933): 444.
8 J. H. Fulton, *The Physiology of the Nervous System* (London: Oxford University Press), 55, citing C. S. Sherrington, "Quantitative Management of Contraction in Lowest Level Coordination," *Brain* 54 (1931): 21.
9 Goldstein, *Der Aufbau des Organismus* (Amsterdam: M. Nijhoff, 1934), chapter 2, 44ff., and chapter 5, 104ff.
10 Cf. Robert Woodworth, *Psychology: A Study of Mental Life* (New York: Henry Holt, 1921), 273, who cites Filimonoff's reports.
11 George E. Coghill, *Anatomy and the Problem of Behavior* (New York: Cambridge University Press, 1929).
12 Cf. Rémy Collin, *L'organisation nerveuse* (Paris: Albin Michel, 1944), 329ff.
13 We can cite, among the numerous cases treated by the Maréville Psychiatric Hospital, that of a female teacher who, after fourteen years of dementia, was able to reprise her job; of a mine engineer who reprised his occupation; of a forest guard, and so forth. We owe these examples to Dr. Hamel.
14 Cf. J. Hermite, *Les mécanismes du cerveau* (Paris: Galliard, 1934), 74ff., which summarizes the observations of Brickner, Dandy, Penfield, and Kleist.
15 [In English in the original.]
16 Nothing is easier than misinterpreting this kind of experiment. Thus, in studying the deferred reaction in monkeys without frontal lobes (for the animal, the task consists in making a correct choice between two pots on the basis of memory), Jacobsen concludes that the ability to record the memory was abolished. But Malmo subsequently found that it was enough to switch off all the lights during the minute in which the animal had to retain the memory (of the pot where the bait is hidden) for the lobotomized animal to carry out the trial successfully. So what is diminished in the animal is not memory but the force of maintaining a conscious assembly despite distractions. Because darkness suppresses every visual distraction, the animal no longer manifests any inferiority. C. F. Jacobsen, "Functions of the Frontal Association Area in Primates," *Archives of Neurology and Psychiatry* 33 (1935):

558–69; Jacobsen, "Studies of Cerebral Functions in Primates: I. The Functions of the Frontal Association Areas in Monkeys," *Comparative Psychology Monographs* 13 (1936): 3–60. For Malmo, see "Interference Factors in Delayed Response in Monkeys after Removal of Frontal Lobes," *Journal of Neurophysiology* 5 (1942): 295–308.

17 Köhler, *Intelligence des singe supérieurs* (Paris: Presses Universitaires de France, 1927), 245.

18 Ruyer, *Éléments de psycho-biologie,* 86ff.

19 Cf. E. Wolff, *La science des monstres* (Paris: Gallimard, 1948), 186.

20 Ruyer, *Éléments de psycho-biologie,* 98.

21 Wolff, *La science des monstres,* 182.

22 Ibid., 239–40.

23 Cf. the works of E. Wolff and V. Dantchakoff.

24 Summarized in Clifford T. Morgan, *Psychological Psychology* (New York: McGraw-Hill, 1943).

25 In the sense that E. Dupréel gave to the word.

26 Paul Weiss, "*Tierisches Verhalten als 'Systemreaktion'*: Die Orientierung der Ruhestellungen von Schmetterlingen (Vanessa) gegen Licht und Schwerkraft," *Biologia Generalis* 1 (1925): 168–248.

27 George Humphrey, *The Nature of Learning* (London: Kegan, Trench, Trubner, 1934).

28 Ibid., 255.

29 Lashley, *Brain Mechanisms and Intelligence,* 141.

30 Humphrey, *Nature of Learning,* 257.

31 [In English in the original.]

32 Cf. H. Prat, *Les gradients histo-physiologiques et l'organogénèse végétale* (Montréal: Laboratoire de botanique, de l'Université, 1945).

33 Wiener, *Cybernetics.*

34 Ruyer, *Éléments de psycho-biologie,* 89.

35 Köhler, *L'intelligence des singes supérieurs,* 170.

36 [In English in the original.]

37 Karl Spencer Lashley, "The Continuity Theory as Applied to Discrimination Learning," *Journal of General Psychology* 26 (1942): 241–65.

38 In his latest work, Köhler, *Gestalt Psychology,* 100–135, also adopts a new theory that is very different from his previous "charged capacitor" models of the brain and, like Lashley, compares a neural incitation arriving from the retina to the striatal area to a circle of waves provoked by a pebble on a pond's surface. Two incitations produce a cortical pattern of interferences. He thus comes close to the thesis previously maintained by Wheeler, another American Gestaltist (cf.

Raymond Holden Wheeler and Francis Theodore Perkins, *Principles of Mental Development* [New York: Crowell, 1932]), who, instead of adopting the theory of cerebral traces, prefers to compare the brain that is receiving several successive excitations to a vibrating plate, like Chladni figures when they are covered with sand (cf. Koffka, *Principles of Gestalt Psychology,* 389–90). The success of these "wave" models of brain function seems to have been favored by the use of "mnemic recorders" in the new calculators.

7. SIGNIFICATION OF EQUIPOTENTIALITY

1 Cf. Ruyer, *Éléments de psycho-biologie,* 82.
2 This is what Samuel Butler vigorously underscores in seemingly humorous expositions. Butler, *Life and Habit* (London: A. C. Fifield, 1910).
3 [A character in Mozart's *The Marriage of Figaro.*]

8. THE RECIPROCAL ILLUSION OF INCARNATION AND "MATERIAL" EXISTENCE

1 Charles Augustus Strong, *Essays on the Natural Origin of the Mind* (London: Macmillan, 1930); Bertrand Russell, *Analysis of Matter* (New York: Harcourt, 1927); Russell, Human Nature (New York: Simon and Schuster, 1948); Eddington, *New Pathways in Science.* Leibniz and the German romantic philosophers had already stated it.
2 We borrow this apt expression from Samuel Alexander, *Space, Time, and Deity: The Gifford Lectures 1916–1918* (London: Macmillan, 1927), without referring to his philosophy (Alexander opposes *enjoyment* and *contemplation*). In an old article ("La Connaissance comme fait cosmique," *Revue philosophique* 113 [1932]: 389–394), we opposed correspondence-knowledge and texture-knowledge; but the term *texture* is equivocal.
3 [In English in the original.]
4 Mark Twain, "Whereas," in *The Works of Mark Twain: Early Tales and Sketches, Volume 2 (1864–1865),* ed. Edgar M. Branch and Robert H. Hirst, 88–83 (Berkeley: University of California Press, 1981).
5 [In English in the original.]
6 Bertrand Russell, *Human Knowledge: Its Scope and Limits* (London: Routledge, 2012), 185.
7 Cf. Tiemen J. C. Gerritsen, *La philosophie de Heymans* (Paris: F. Alcan, 1938), 247ff.
8 The expression comes from Heidegger, but the thesis is not specifically

existentialist or Hegelian. Cf., e.g., N. Hartmann, *Ethik* (Berlin: W. de Gruyter, 1926), 312.

9 Cf. Alexis Moyse, *Biologie et physico-chimie* (Paris: Presses Universitaires de France, 1948), 77.

10 G. Matisse, *Le rameau vivant du monde* (Paris: Presses Universitaires de France, 1947), 3:217.

11 Giovanni Malfitano and M. Catoire, *Introduction à la chimie micellaire* (Paris: Hermann, 1942).

12 Georges Lemaître, *The Primeval Atom: An Essay on Cosmogony* (New York: Van Nostrand, 1950).

9. "ABSOLUTE SURFACES" AND ABSOLUTE DOMAINS OF SURVEY

1 Cf. Raymond Ruyer, *La conscience et le corps* (Paris: F. Alcan, 1950), 56ff., and Ruyer, "Sur une illusion dans le théories philosophiques de l'étendue," *Revue de Métaphysique* 4 (1933): 521–27.

2 J. W. Dunne, *The Serial Universe* (London: Faber and Faber, 1940), 29ff.

3 Ibid., 32.

4 Ibid.

5 Chapter 6.

6 Chapter 8 and Ruyer, *La conscience et le corps,* part I.

7 Chapter 6.

8 A. Wenzl glimpsed this necessity in *Wissenschaft und Weltanschauung,* part III (Leipzig, Germany: M. Meiner, 1936). But he mistakenly inferred from it the necessity of admitting a fifth dimension, which amounts to conflating geometry and consciousness.

10. ABSOLUTE DOMAINS AND BONDS

1 Cf. Louis de Broglie, *Physique et microphysique* (Paris: Albin Michel, 1947), 161; and Henri Bouchet, *Introduction à la philosophie de l'individu* (Paris: Flammarion, 1949), 39.

2 Cf. chapter 18.

3 Cf. Wolff, *La science des monstres,* 32.

4 [In English in the original.]

11. ABSOLUTE DOMAINS AND FINALITY

1 Raymond Aron, *Introduction à la philosophie de l'histoire: essai sur les limites de l'objectivité historique* (Paris: Gallimard, 1978).

2 A. A. Cournot, *Considérations sur la Marche des Idées et des Evènements dans les Temps Modernes,* ed. F. Mentré (Paris: Boivin et Cie, 1934), 1–2.
3 According to Aron, *Introduction à la philosophie de l'histoire,* 164.
4 Cf. chapter 2 and Ruyer, "Métaphysique du travail."
5 Wiener, *Cybernetics,* 34.
6 David Hume, *Dialogues Concerning Natural Religion and Other Writings,* ed. Dorothy Coleman (Cambridge: Cambridge University Press, 2007), 37.
7 Ibid., 38.

12. THE REGION OF THE TRANSSPATIAL AND THE TRANSINDIVIDUAL

1 Henri Pieron, "Recherches comparative sur la mémoire des formes et celle des chiffres," *L'année psychologiques* 21 (1920–21): 119–48.
2 Robert Sessions Woodworth and Harold Schlosberg, *Experimental Psychology* (New York: Holt, Reinhart, and Winston, 1954), 24.
3 R. Goldschmidt, "The Theory of the Gene," *Scientific Monthly* 45 (1938): 268–73.
4 Robin George Collingwood, *The Idea of Nature* (Oxford: Clarendon Press, 1945), 146.

13. THE LEVELS OF THE TRANSSPATIAL AND FINALIST ACTIVITY

1 [In English in the original.]
2 Abel Hermant, *Amour de tête* (Paris: G. Charpentier, 1890).
3 Kantor, *Problems of Physiological Psychology,* 105.

14. THE BEINGS OF THE PHYSICAL WORLD AND THE FIBROUS STRUCTURE OF THE UNIVERSE

1 Nicolai Hartmann, *New Ways of Ontology* (Chicago: H. Regnery, 1953), 87.
2 Isaac Newton, *Philosophical Writings,* ed. Andrew Janiak (Cambridge: Cambridge University Press, 2004), 87–88.
3 Cf. Bierens de Haan, *Animal Psychology for Biologists* (London: University of London Press, 1929), 29.
4 Louis de Broglie, *Continu et discontinu en physique moderne* (Paris: Albin Michel), 66, 74.
5 E. T. Whittaker, *Space and Spirit* (London: Thomas Nelson, 1946), 113–14.

6 We have already noted that it is "action" and not "work," in the sense that these terms have in the lexicon of physics, that corresponds to work-activity in the ordinary sense.
7 De Broglie, *Continu et discontinu,* 30ff.
8 Ibid., 36.
9 L. Susan Stebbing, *Philosophy and the Physicists* (London: Taylor and Francis, 1969), 275.
10 James Horne Morrison, *Christian Faith and Science* (Nashville, Ill.: Cokesbury Press, 1936), 212.
11 Stebbing, *Philosophy and the Physicists,* 275.
12 Gaston Bachelard, *The New Scientific Spirit* (Boston: Beacon Press, 1984), 62–63.
13 Collingwood, *Idea of Nature,* 146.
14 Ibid., 147.
15 Ibid., 148.
16 Cf. Ruyer, "Métaphysique du travail," 208ff.
17 The first manifestation of this "substantialized" memory is perhaps the phenomenon of protein viruses' reproduction by self-replication. The chemical operations of living beings, which pass from unstable forms to unstable forms in very complex chain reactions, of which we grasp only a few stages and a few snapshots, have already something of a hereditary mnemic melody. Alexis Moyse insisted on the very interesting idea of a melodic continuity of unstable forms within chemical reactions in biology: "The realization of these forms is so fleeting that it eludes us; we cannot seize them, capture them, although we believe that we are entitled to suppose their existence (e.g., the formaldehyde in the chlorophyllic synthesis; oxygenated water in the final phase of respiratory oxidation). . . . Our intervention in the study of these mechanisms is comparable to that of a watchmaker who is forced from time to time to block the needles of his watch in order to read the time." Moyse, *Biologie et physico-chimie* (Paris: Presses Universitaires de Paris, 1948), 34.
18 Georges Lemaître, *The Primeval Atom: An Essay on Cosmogony* (New York: Van Nostrand, 1950).
19 [In English in the original.]
20 Cf. Paul Couderc, *The Expansion of the Universe* (London: Macmillan, 1952), and Fred Hoyle, *The Nature of the Universe* (Oxford: Basil Blackwell, 1951), 46ff.

15. THE NEOMATERIALIST THEORIES

1 The "dynamist" conceptions of the organism, which turns it into a set of equilibria or stationary processes akin to the equilibria of macroscopic physics, are no doubt of the same order and fare no better (cf. for example the laborious developments of W. Köhler, who continues to defend such conceptions in *The Place of Values in a World of Facts* [New York: Liveright, 1938], chapter VIII).

2 Except for a few "purists," such as E. Rabaud and L. Hogben.

3 Cf. on this movement William MacDougall, *The Riddle of Life* (London: Methuen, 1938), 97ff.

4 Benjamin Moore, *The Origin and Nature of Life* (New York: Henry Holt, 1913).

5 Max Loewenthal, *Life and Soul: Outlines of a Future Theoretical Physiology and of a Critical Philosophy* (New York: Allen and Unwin, 1934).

6 Augusta Gaskell, *What Is Life?*, with prefaces by K. T. Campton and R. Pearl (Springfield, Ill.: C. C. Thomas, 1928). We know this work only through MacDougall's summary in *Riddle of Life*, 113.

7 Niels Bohr, *Atomic Theory and the Description of Nature* (Cambridge: Cambridge University Press, 1934).

8 N. Bohr's idea is inaccurate: we can perfectly experiment on an organism in its specificity. Consider the experimental grafts of embryology, the experiments of electric excitations of the cortex of a nonanesthetized patient, and so forth. It is true that we cannot "observe" the psychobiological as such, but this law falls within the more general law of the inobservability of bonds.

9 Niels Bohr, "Light and Life," *Nature* 131 (1933): 421–23.

10 Ralph S. Lille, *General Biology and Philosophy of Organism* (Chicago: University of Chicago Press, 1946), esp. chapters 4 and 9.

11 Jan Christiaan Smuts, *Holism and Evolution* (London: Macmillan, 1936).

12 Despite certain declarations, for example, Moyse, *Biologie et physico-chimie*, 66–67.

13 Georges Matisse, *Le rameau vivant du monde* (Paris: Presses Universitaires de France, 1947), 3:16.

14 P. Jordan, *Anschaulische Quantentheorie: Eine Einführung in die modern Auffassing der Quantenerscheinungen* (Berlin: Springer, 1936).

15 Erwin Schrödinger, *What Is Life?* (Cambridge: Cambridge University Press, 2012), 4. Recall that, for modern physics, a molecule, a crystal, and a true solid are not genuinely different. They are opposed in their

entirety to amorphous states: gas, liquid, or pseudo-liquid (noncrystallized solids).

16 Ibid., 5.
17 Ibid., 21.
18 Ibid., 55.
19 Ibid., 73.
20 Harold F. Blum, in a book that appeared during the printing of the current work, *Time's Arrow and Evolution* (Princeton, N.J.: Princeton University Press, 1951), sustains a neomaterialist conception very close to E. Schrödinger's.
21 Schrödinger, *What Is Life?*, 82.
22 Ibid., 77.
23 Ibid., 88.

16. NEO-DARWINISM AND NATURAL SELECTION

1 We take Democritus as a simple figurehead for a doctrinal schema. The real Democritus is much more complex. In his physics, he seems to shift quickly from the reign of pure chance to the reign of laws. What he calls "sorting" *(diacrisis)* is not a pure sorting of fortuitous combinations—this only takes place at the outset *(peripalaxis)*—but a regular, "oriented" sorting, as G. Matisse would say, like a centrifugation, the sifting of wheat, or a sieving. There are fewer differences than is typically thought between Democritean and Cartesian atomism. In reality, nearly all the believers in chance and in "motor causes" appeal to atemporal geometric laws, which impose formal conditions of existence on the would-be fortuitous products of shocks and of motor causes.
2 W. Ludwig's comparison.
3 Cyril Darlington, *The Evolution of Genetic Systems* (London: Oliver and Boy, 1958).
4 William Dampier, *A History of Science and Its Relations with Philosophy and Religion* (London: Cambridge University Press, 1971), 281. Cf. also S. Butler, "The Attempt to Eliminate Mind," chapter 10 in *Luck or Cunning* (London: A. C. Fifield, 1920).
5 Ronald Aylmer Fisher, *The Genetical Theory of Natural Selection* (Oxford: Clarendon Press, 1930). A good exposition of S. Wright's mathematical calculations is provided by W. Ludwig, "Die Selektiontheorie," in *Die Evolution der Organismen,* ed. G. Heberer (Jena, Germany: Fischer, 1943), 497ff.
6 Such is the attitude of Rabaud and his disciples in France.

7 J. Huxley, *Evolution: The Modern Synthesis* (Boston: MIT Press, 2009), 412.
8 Ibid., 473. [In English in the original.]
9 Ibid.
10 Ibid., 387.
11 Cf. ibid., 56.
12 If selection is supposed to discriminate between two mutants whose degrees of fitness only differ in the one hundredth or the one thousandth, how can it let races and species with monstrously hypertelic or dystelic organs subsist? Neo-Darwinism is forced to espouse two contradictory politics: at times, selection is an instrument of infinite sensitivity that discriminates between mutants whose differences are imperceptible; at others, it is singularly liberal or rough. The neo-Darwinian explanations of the facts of dystelia (intraspecific selection, connection with favorable traits, and so forth) are an auxiliary hypothesis generated to sustain a theory rather than to docilely interpret the facts.
13 Wiener, *Cybernetics,* 34.
14 Cott, *Adaptive Colorations in Animals,* 72.
15 Stephenson and Stewart, *Animal Camouflage,* 108.
16 Thomas Hunt Morgan, *Evolution and Genetics* (Princeton, N.J.: Princeton University Press, 1925), 148–50.
17 Gavin de Beer, *Embryology and Evolution* (Oxford: Clarendon Press, 1930).
18 Huxley, *Evolution,* 499.
19 Cf. Charles Rupert Stockard, *The Physical Basis of Personality* (New York: W. W. Norton, 1931).
20 Darlington, *Evolution of Genetic Systems,* 46.

17. NEO-DARWINISM AND GENETICS

1 [Ruyer could not have anticipated Watson and Crick's pivotal discovery of the structure of DNA in 1953, one year after the publication of *Neofinalism.* For an excellent analysis of modern Darwinism and its genetic basis, see Stephen Jay Gould, *The Structure of Evolutionary Theory* (Cambridge, Mass.: Harvard University Press, 2002).]
2 We summarize them quickly here, because we have analyzed them at length in our previous work *Élément de psycho-biologie,* chapters 3 and 8.
3 Pensfield and Boldrey. Morgan, *Physiological Psychology,* reproduces their schema of the cortical homunculus.
4 Morgan, *Embryology and Genetics,* 129.

5 Ibid.

6 Marcel Prenant, *Biologie et marxisme* (Paris: Éditions Hier et Aujourd'hui, 1948), 183.

7 Ibid., 184.

8 Cf. de Beer, *Embryology and Evolution,* and Gavin de Beer, *Embryos and Ancestors* (Oxford: Clarendon Press, 1940).

9 Of E. Wolff and of V. Dantchakoff.

10 P. Ancel and his collaborators (cf. Ancel, *Chimiotératogénèse* [Paris: G. Doin, 1950]) have shown that many of the monstrosities that can be produced experimentally (by introducing viruses or chemicals into the embryo) faithfully reproduce hereditary monstrosities and monstrosities with a germinal origin. This is the sign that hereditary monstrosities are also due to the production of chemicals.

11 C. C. Hurt, *Heredity and the Ascent of Man* (London: Cambridge University Press, 1935), vii.

12 [In English in the original.]

13 Cf. on this subject Fisher, *Genetical Theory of Natural Selection*; Cott, *Adaptive Colorations in Animals,* 423; Huxley, *Evolution*; and Lucien Chopard, *Le mimétisme* (Paris: Payot, 1949).

14 Cf. Chopard, *Le mimétisme,* 317.

15 J. M. Baldwin, *Development and Evolution* (New York: Macmillan, 1902).

18. ORGANICISM AND THE DYNAMISM OF FINALITY

1 Organicism remains at the classical conception of physics and generally does not appeal to contemporary microphysics. Smuts is an exception; in his latest work, he combines "holism" with a conception very close to Lillie's.

2 For example, here is a typical passage from K. Goldstein: "Wir suchen nicht einen Realgrund, des Sein begründet, sondern eine Idee, den Erkenntnisgrund, in dem alle Einzelheiten ihre Bewährung erfahren, eine 'Idee', von der aus all die Einzelheiten verständlich warden, wenn wir die Bedingungen ihrer Entschehung berücksichtigen." Goldstein, *Der Aufbau des Organismus* (The Hague: M. Nijhoff, 1934), 242. The chiaroscuro of this type of text requires its original language.

3 Cf. Louis Bounoure, *L'autonomie de l'être vivant* (Paris: Presses Universitaires de France, 1949), 202.

4 René Descartes, *Philosophical Writings of Descartes,* trans. J. Cottingham, R. Stoothoff, and D. Murdoch (Cambridge: Cambridge University Press, 1985), 1:99.

5 See esp. Immanuel Kant, *Critique of Judgment* (Cambridge: Cambridge University Press, 2000), sections 67 and 78, and the preface to Kant, *Universal Natural History and Theory of the Heavens* (Ann Arbor: University of Michigan Press, 1969).
6 Kant, *Critique of Judgment,* 251.
7 Ibid.
8 This is still the thesis of Bernard Bosanquet, *The Meaning of Teleology* (London: Henry Frowde, 1906), and of Lawrence Joseph Henderson, *The Fitness of the Environment* (New York: Macmillan, 1913).
9 G. Matisse summarized them in *Le rameau vivant du monde* (Paris: Presses Universitaires de France, 1948), 3:95ff.
10 Claude Bernard, *An Introduction to the Study of Experimental Medicine* (New York: Dover, 1957), 93.
11 Claude Bernard, *Lectures on the Phenomena of Life Common to Animals and Plants* (Springfield, Ill.: C. C. Thomas, 1974), 1:37.
12 Ibid., 1:38.
13 Ibid., 1:34.
14 Ibid., 1:240.
15 It is fair to say that the same kinds of hesitations can be found among avowed vitalists like Driesch or Reinke. Driesch speaks of his entelechy as an "agent" but also declares that it is not "a sort of energy." To reconcile the irreconcilable, he admits that it can suspend the conversion of potential energy into kinetic energy but cannot control this conversion and that the process of suspension itself does not require energy (which is a plain scientific error). Hans Driesch, *Philosophie de l'organisme* (Paris: M. Rivière, 1921), 2:221. The excuse of vitalists and of organicists is that the problem was insoluble before the rise of microphysics.
16 *The Organismal Conception,* quoted in MacDougall, *Riddle of Life,* 151–52.
17 Ibid., 152.
18 Ibid., 151.
19 Ludwig von Bertalanffy, *Kritische Theorie der Formbildung* (Berlin: Gebrüder Borntraeger, 1928).
20 Bounoure, *L'autonomie de l'être vivant,* 212, 215.
21 Ibid., 209.
22 Ibid., 214.
23 Goldstein, *Der Aufbau des Organismus.*
24 Ibid., chapters 2 and 5.
25 Ibid., chapter 6.
26 Ibid., 141.

27 Ibid., 145.
28 Ibid., 169.
29 Ibid., 261.
30 Ewald Oldekop, *Le principe de hiéarchie dans la nature* (Paris: Vrin, 1933).
31 Goldstein, *Der Aufbau des Organismus,* 263.
32 Ibid., 257.
33 Edward Chace Tolman, *Purposive Behavior in Animals and Men* (Berkeley: University of California Press, 1949), 7.
34 Ibid.
35 Ibid., 418.
36 [In English in the original.]
37 [In English in the original.]
38 Maurice Merleau-Ponty, *The Structure of Behavior* (Boston: Beacon Press, 1963), 223n.
39 Ibid., 143.
40 Cf. Ruyer, *La conscience et le corps,* 10ff.
41 J. J. von Uexküll, "Der Organismus und die Umwelt," in *Das Lebensproblem im Lichte der modernen Forschung,* ed. H. Driesch and H. Woltereck (Leipzig, Germany: Quelle und Meyer), 223. Let us note that von Uexküll is Kantian in his general philosophy (cf. *Theoretical Biology* [New York: Harcourt, Brace, 1926], preface) and conflates, like Merleau-Ponty, comprehensive biology and critical biology. For example: "All reality is subjective appearance. This must constitute the great, fundamental admission even of biology" (xv). We borrow his phrase without referring to his general doctrine.
42 Merleau-Ponty, *Structure of Behavior,* 159.
43 De Broglie, *Revue de synthèse,* 1934 [citation is incomplete].
44 Anntoine Cournot, *Traité,* vol. 2 (Paris: Hachette, 1861), chapter 6. We should equally mention Charles Peirce here.
45 Ibid., section 135.
46 Ibid., section 139.
47 W. Heitler, *Elementary Wave Mechanics* (London: Oxford University Press, 1946), 95–96.
48 Ibid.
49 [In English in the original.]
50 Heitler, *Elementary Wave Mechanics,* 112.
51 Merleau-Ponty, *Structure of Behavior,* 242n35.
52 We drew the idea for this myth from a philosophical work that we have since lost.
53 Butler, *Notebooks of Samuel Butler,* 336.

54 Boonoure, *L'autonomie de l'être vivant,* 216.

55 Arthur Lovejoy, *The Revolt against Dualism* (La Salle, Ill.: Open Court, 1955), chapter 2. Cf. also the collection Arthur Lovejoy et al., *Essays in Critical Realism* (New York: Macmillan, 1920).

56 Ruyer, *Conscience et le corps.*

57 Cf. Merleau-Ponty, *Structure of Behavior.* We find a similar confusion in Lossky concerning the Bergsonian theory of perception; cf. Ruyer, *Le monde des valeurs,* 162ff.

58 Cf. Köhler, *Place of Values in a World of Facts,* chapter 9. Note that there is no need to reproach Köhler for *comparing* the activity that aims for a value with physical force. We made this same comparison ourselves before stumbling on his work (cf. Ruyer, *Éléments de psychobiologie,* 266). What he should be reproached for is having compared axiological activity with the action of a force of macroscopic physics. Thus, although he denies this, he reduces axiological and finalist action to a causal influence that establishes itself step by step and culminates in a "molar" equilibrium.

59 Samuel Butler considered the digestion and assimilation of food as acts of "proselytism." This conception, like the majority of Butler's humorous ideas, is literally true.

19. PSYCHO-LAMARCKISM

1 Paul Wintrebert, "Le Lamarckisme chimique," *Comptes rendus de l'Académie des sciences de Paris* 228 (1949): 1079–82, spoke of a "chemical Lamarckism." One race of microbes can noticeably acquire the traits of a related race through the action of a nucleic acid (chemical induction, guided mutation). But it is best not to speak of Lamarckism in such cases.

2 G. G. Simpson, *Tempo and Mode in Evolution* (New York: Columbia University Press, 1984), 180.

3 Henri Bergson, *Creative Evolution* (New York: Henry Holt, 1911), 181.

4 Ibid., 78.

5 Bleuler is the most interesting of the three. See esp. Eugen Bleuler, *Mecanismus, Vitalismus, Mnemismus* (Berlin: Springer, 1931).

6 H. Driesch, *Philosophie de l'organisme* (Paris: Rivière, 1921), 127.

7 Ibid., 222n.

8 Cf. J. Huxley, *Soviet Genetics and World Science: Lysenko and the Meaning of Heredity* (London: Chatto and Windus, 1949).

9 Cf. Huxley, *Evolution,* and Simpson, *Tempo and Mode in Evolution.*

10 Carl Gustav Carus, *Psyche: On the Development of the Soul* (New York: Spring, 1970).
11 See the preface to George Bernard Shaw, *Back to Methuselah* (New York: Penguin Books, 1988).
12 [In English in the original.]
13 Butler, *Life and Habit,* 61.
14 Ibid.
15 [In English in the original.]
16 Pierre Jean, *Dieu or la physique* (Paris: R.-A. Corrêa, 1935), 60.
17 J. Holmes, *The Problem of Organic Forms,* cited by MacDougall, *Riddle of Life,* 226.
18 MacDougall quite rightly points this out.
19 Jean, *Dieu ou la physique,* 51.
20 [In English in the original.]
21 Goblot's theory is the most well known.
22 [In English in the original.]
23 Cournot, *Traité,* section 227.
24 Simpson, *Tempo and Mode in Evolution,* 187.
25 [In English in the original.]
26 Simpson, *Tempo and Mode in Evolution,* 206.
27 Ibid.
28 Cf. Lucien Cuénot, *L'adaptation* (Paris: Gaston Doin, 1925).
29 Samuel Butler, *God the Known and God the Unknown* (London: A. C. Fifield, 1926). Burloud's remarkable book *De la psychologie à la philosophie,* which we have not considered in this discussion, moves very significantly from psychologism to theism. Albert Burloud, *De la psychologie à la philosophie* (Paris: Hachette, 1950).

20. THEOLOGY OF FINALITY

1 Adolphe Lods, *Histoire de la littérature hébraïque et juive* (Paris: Payot, 1950), 935.
2 Cf. Ruyer, *Le monde des valeurs,* chapter 8.
3 Von Uexküll, *Theoretical Biology,* 138.
4 Alexander, *Space, Time, and Deity,* 2:347. For Alexander and organicists, a being of a given level can be entirely described, without residue, in the terms of the lower level; the *novum* is just the quality that constitutes the *soul* of the configuration realized by the unities of the lower level or its new "color." Deity is the "color" that the universe will assume.
5 Alexander's theory would be valid if he limited himself to saying that,

for us, "deity" can be understood as founded on personal life, although not definable according to the category of personality.

6 Alexander, *Space, Time, and Deity,* 398.
7 N. Hartmann, *Ethics* (New York: Macmillan, 1932), 1:285ff.
8 Ibid., 290.
9 Lawrence J. Henderson, *The Order of Nature* (Boston: Harvard University Press, 1917), 5. Like Bosanquet, Henderson adopts the Kantian thesis of the *Critique of Judgment,* but the value of his arguments is not contingent on this thesis.
10 Ibid., 184.
11 Cf. Philippe Olmer, *La structure des choses* (Paris: Hachette, 1949), 202–3.
12 Henderson, *Order of Nature,* 186.
13 Ibid., 192. The recent work of the biochemist Harold F. Blum, *Time's Arrow and Evolution* (Princeton, N.J.: Princeton University Press, 1951), while remaining neomaterialist and antifinalist, undertakes a curious synthesis of E. Schrödinger and of Henderson. Blum too thinks that adaptation presupposes the fitness of the physicochemical world, which thus channels evolution. He adds a whole host of clarifications to Henderson's thesis, especially on the fitness of hydrogen.
14 Hume, *Dialogues Concerning Natural Religion,* 54.
15 Ibid., 55, 54.
16 Cf. Marcel Boll, *Les deux infinis* (Paris: Larousse, 1938), 216.
17 Georges Matisse, "Le hasard et les phénomènes orientés," *Revue de Métaphysique et de Morale,* no. 1 (1938): 9–34.
18 Alfred North Whitehead, *Process and Reality* (New York: Free Press, 1978), 7; cf. also 40ff. and the last chapter of the work.
19 André Lalande, *La raison et les norms* (Paris: Hachette, 1948), 12.

TRANSLATOR'S AFTERWORD

This talk was delivered in 2008 to preface a series of seminars on "anoialogy," the logic of thought. The experimentation and ideas of those feverish weeks have been preserved in a forthcoming volume, *Aporia and Philosophy: Seminars.*

1 "To gar auto noein estin te kai einai" (Parmenides, fragment 3).
2 It is certainly no stroke of luck that the knot of Being and Thinking is inscribed in philosophy's lexicon as well as in its fate. According to Varro, *cogitare,* "to think," derives from *cogere,* meaning "to gather, to collect, to bring together"; from the start, thought is esteemed to be a gathering, a herding of Being. See Varro, *De Lingua Latina,* VI

43 (Cogitare a cogendo dictum: mens plura in unum cogit, unde eligere possit). And it is to this knot that the latest *diadoche* of the Parmenidean cry addressed his grandiose eulogy. See Martin Heidegger, "*Moira*," *Vorträge und Aufsätaze* (Pfullingen, Germany: Neske, 1967), 3:231–56.

3 Cf. Averroës's asubjective topology: "[al-'akl] dākhil 'alayna min [al] khārij" (thought penetrates us from the Outside). *Tafsir ma ba'd at-tabi'at*, ed. Maurice Bouyges (Beirut: Imprimerie Catholique, 1948), 3:1489.

Index

(continued from page ii)

12 A FORAY INTO THE WORLDS OF ANIMALS AND HUMANS, *WITH* A THEORY OF MEANING
Jakob von Uexküll

11 INSECT MEDIA: AN ARCHAEOLOGY OF ANIMALS AND TECHNOLOGY
Jussi Parikka

10 COSMOPOLITICS II
Isabelle Stengers

9 COSMOPOLITICS I
Isabelle Stengers

8 WHAT IS POSTHUMANISM?
Cary Wolfe

7 POLITICAL AFFECT: CONNECTING THE SOCIAL AND THE SOMATIC
John Protevi

6 ANIMAL CAPITAL: RENDERING LIFE IN BIOPOLITICAL TIMES
Nicole Shukin

5 DORSALITY: THINKING BACK THROUGH TECHNOLOGY AND POLITICS
David Wills

4 BÍOS: BIOPOLITICS AND PHILOSOPHY
Roberto Esposito

3 WHEN SPECIES MEET
Donna J. Haraway

2 THE POETICS OF DNA
Judith Roof

1 THE PARASITE
Michel Serres

RAYMOND RUYER (1902–1987) was a professor of philosophy at the Université de Nancy. His interests ranged from metaphysics to the philosophy of science and Greek mythology. He was the author of more than twenty books, including *Elements of Psychobiology* (*Éléments de psycho-biologie,* 1946), *The Genesis of Living Forms* (*La genèse des formes vivantes,* 1956), and *Cybernetics and the Origin of Information* (*La cybernétique et l'origine de l'information,* 1954).

ALYOSHA EDLEBI is the editor of *What Is Thinking?* (forthcoming).

MARK B. N. HANSEN is professor of literature at Duke University. He is the author of *Bodies in Code: Interfaces with New Media* and *New Philosophy for New Media.*